Freedom and Self-Creation

Freedom and Self-Creation

Anselmian Libertarianism

Katherin A. Rogers

OXFORD
UNIVERSITY PRESS

Great Clarendon Street, Oxford, OX2 6DP,
United Kingdom

Oxford University Press is a department of the University of Oxford.
It furthers the University's objective of excellence in research, scholarship,
and education by publishing worldwide. Oxford is a registered trade mark of
Oxford University Press in the UK and in certain other countries

First Edition published in 2015

Impression: 1

Published in the United States of America by Oxford University Press
198 Madison Avenue, New York, NY 10016, United States of America

British Library Cataloguing in Publication Data
Data available

Library of Congress Control Number: 2015934140

ISBN 978–0–19–874397–2

Printed and bound by
CPI Group (UK) Ltd, Croydon, CR0 4YY

Contents

Introduction: Anselmian Libertarianism

> [The good] angels are not to be praised for their justice due to the fact that they were able to sin, but rather due to the fact that, in a way, they have it from themselves that they are [now] unable to sin; in this they are, to some extent, similar to God, who has whatever He has from Himself (*a se*).
>
> *Cur deus homo* 2.10[1]

Part I: Why Do We Want Free Will?

Introduction

Anselm of Canterbury is probably the first philosopher on the planet to attempt a well-worked-out theory of libertarian free will.[2] He is motivated to pioneer this position by the thought that if we human beings (and any other created rational and free agents there may be) are to bear any responsibility for what we do and, more importantly, for the kind of people we are, then we must be able to make choices which truly come from ourselves. He believes, as an entailment of his Christian faith, that God made us in His image. And he takes that to include the point that we are remarkable reflections in that we are able to participate in the creation of our own characters. The purpose of freedom is that we should be able to imitate God, not just by being good creatures—any creature is a good creature—but by being good *from ourselves*. We exist in absolute dependence on God, but He has opened for us a small space for independent action and self-creation. The locus of that small space is free choice.

The core of Anselm's theory is that the created free agent must be able to choose between open options and that choice must come in some ultimate way

[1] Translations throughout are my own from the standard Latin text, *Anselmi Opera Omnia*, ed. F.S. Schmitt, 6 vols. (Rome and Edinburgh: Friedrich Frommann Verlag, 1938–68).

[2] See my *Anselm on Freedom* (Oxford: Oxford University Press, 2008).

from himself.[3] He is what we, today, would refer to as an agent-causalist, but with the difference that he insists that we must not attribute to the created free agent any new and unique causal powers. This proposal of a *parsimonious* agent-causation constitutes a novel and exciting contribution to the free will discussion, but it seems to have been lost to intellectual history.[4] My aim in the present work is to recapture the theory as a viable approach within the current free will debate. I begin (Chapter One) with a defense of Anselm's basic motivation; if God causes absolutely everything, including the choices of created agents, then those agents are not free, are not responsible, and cannot be the *imagines dei* that Anselm takes human beings to be. I attempt to cast the discussion in the contemporary idiom and within the contemporary literature, and so a new character, the Anselmian, has to be introduced. The Anselmian embraces the basic outline initially proposed by Anselm, and then attempts to fill it in and build upon it. "Anselm" and "Anselm's" indicate that it is indeed the historical figure whose thought is expressed. As Chapter One exemplifies, Anselm's theory translates well into the present debate. In Chapter One Anselm's insistence that we cannot be free if all our choices are caused by God provides a stronger version of contemporary "controller" arguments for incompatibilism. In Chapters Two through Four I spell out Anselm's parsimonious agent-causation and its entailments with an eye to how his theory connects with the contemporary debate.

Anselm's theory is, of course, a proper target for the usual criticisms that can be leveled against libertarianism in general, and agent-causal libertarianism in particular. Sometimes elements in his theory—like his refusal to hypothesize *sui generis* causation unique to free choice—allow it to escape, or at least significantly mitigate, some of the standard problems. But sometimes elements in his theory—such as his denial that choices are robust "things" with ontological status—make some of the standard problems seem, prima facie, more obvious or more damning. Chapters Five through Eight defend Anselmian libertarianism by proposing

[3] Anselm is what Kevin Timpe would label a "wide source incompatibilist": the core of free will is that the agent be the ultimate source of his choices, and that entails—for the created agent, at least—that he confront alternative possibilities; *Free Will: Sourcehood and Its Alternatives* (London: Continuum, 2008).

[4] To my knowledge, Anselm's theory of free will was not adopted in toto by any subsequent thinkers. Duns Scotus is a libertarian and was influenced by Anselm, but his overall theory does not seem to be the same (though my knowledge of Scotus is limited, and perhaps he is closer to Anselm than I take him to be). Among the historical reasons for this "forgetfulness" we might list the Western European reception of Aristotle towards the end of the twelfth century. Anselm's *Proslogion* argument and his *Cur deus homo* were well known to thirteenth-century thinkers, as shown by Aquinas' criticisms of both, but perhaps the details of Anselm's theory of freedom got lost in the excitement over Aristotle.

Anselmian answers (answers drawn from, or inspired by, Anselm's theory) to three of the most significant and standard problems raised against libertarianism. (I offer four chapters in response to three problems because the "luck" problem requires two chapters of Anselmian response.) Chapter Nine raises, and attempts to answer, a question that has not seen a great deal of play in the current literature, but which is especially acute for the Anselmian theory: The point of freedom is self-creation, but if agents are not aware of this, can they be held responsible for the selves they create? Chapter Nine ends with a brief recapitulation of, and conclusion to, the book as a whole.

Before we can move to Anselm's theory and its' motivations some introductory material is in order. This introductory chapter is divided into two parts. The second part provides a road map to the book as a whole. In this first part I begin by setting out an important methodological principle that will be in play throughout the book: Anselm's theory is to be explained and defended with an eye to Anselm's own theist perspective. This is a constructive way to go about things for two reasons: First, showing how Anselm's (and the Anselmian's) position on free will arises within his theism illustrates a way in which background beliefs may play a significant role in generating intuitions concerning various steps in the free will debate. And second, it is helpful, in a number of arguments, to invoke an ideal knower and/or an omnipotent controller. In Anselm's universe, God is already there to fill these roles.

I continue Part I of the introductory material by explaining how I will be using the terms relevant to the discussion. Especially important will be two sorts of responsibility-denying necessity: causal necessity and "external, non-causal" necessity, as distinguished from "consequent" necessity which is entirely innocuous vis-à-vis freedom and responsibility. Next I note that science and experience leave open the question of what sort of free choices we make, if any. The absence of any conclusive evidence one way or the other leaves the philosopher free to ponder the issue of why we *want* free will. Anselm, and those of a like mind, want free will—a very robust, libertarian free will—because it is required (or so I will argue) if we are to be the metaphysically valuable things that Anselm, and much of Western thought, has traditionally held us to be. Part I of the Introduction concludes with setting out still more methodological principles. First, it is choices, and not overt deeds that will be the focus of consideration. Second, in that the Anselmian connects freedom with the value of the free agent, criteria that are inclusive are, *ceteris paribus*, preferable to criteria that are exclusive in describing what is required for free choice. And finally, it needs to be noted up front that the concern is less with individual choices and acts than with the character that is built through those choices and acts. As I noted above, in

Anselm's view the value of free will is that it allows for self-creation. This point will both solve and generate problems towards the end of this book.

Worldviews and Intuitions

In *De casu diaboli*, where Anselm begins to work out the details of his libertarian theory, he makes the central assumption that if the situation has been engineered by God such that the created agent can "choose" only one way, then the agent chooses by necessity and is not choosing freely.[5] I take this to mean that Anselm shares the basic libertarian intuition that if your choice is causally necessitated by something outside of you and not "up to" you, then your choice was not free. I will refer to this as the "basic intuition" and try to develop and defend it in Chapter One. It should be noted that Anselm himself offers what could be an argument—or a premise in an argument—in favor of libertarianism when he insists, in *De libertati arbitrii* 8, that it is *logically impossible* that God cause sin, and yet sin happens. It follows—in Anselm's universe—that it is entirely up to the created agent to cause it.[6] But this argument is persuasive only if one supposes that the only sort of causal necessity that might conflict with free will comes from God. Anselm himself was not concerned about other causes that might necessitate an agent's choice, and, in his universe, any natural necessitating causes would ultimately trace their efficacy back to God in any case. The Anselmian, though, is trying to bring Anselm's theory into dialogue with contemporary philosophers, and so must be concerned about any and all necessitating causes, divine or earthly. So I will treat Anselm's fundamental motive for developing his libertarianism as deriving from the "basic intuition." Anselm does seem to "just assume" it in many contexts, and, in that intuition drives so much of the contemporary debate, I believe this will be the most helpful way to proceed. It will be well here in the Introduction to say a little about intuition and how intuition will be employed in the present work.

An advantage of referring to "the Anselmian" (rather than just noting or implying that certain views are my own) is that it will serve to remind the reader that a large and well-worked-out theistic worldview serves as the backdrop and the impetus for Anselmian libertarianism. This is not to say that the theory will be of interest only to those who share the background beliefs. I believe anyone interested in free will issues should find the theory, its entailments, and its responses to problems worthy of attention. But sometimes it can be helpful for participants in philosophical debates to lay all their cards on the table. I want to

[5] *De casu diaboli* (DCD) 13–14. [6] Rogers (2008): 89.

do that here, because I believe it will be useful to underline the connection between background beliefs and intuitions.

I will *mean* by "intuition" what I take most of us engaged in the free will discussion to mean: roughly an intellectual "seeming" or appearance that some proposition is true or some state of affairs obtains, which seeming is not the product of sense perception or other experiential evidence, but derives from an understanding of the proposition or state of affairs. For example, many philosophers, upon considering the basic description of libertarian freedom, find that such freedom involves an unacceptable element of luck. Their "finding" it so is based on an intellectual sort of "just seeing" that it is the case. I have no qualms about allowing intuitions as prima facie evidence for the truth of what is intuited.[7] But when intuitions collide one searches for ways to encourage the opponent to see what one sees oneself. Or, failing that, one tries to clarify the opposing intuitions.

Anselm himself does not discuss intuition per se. Doubtless he would agree with Augustine's claim that there are necessary truths that all of us "just see"; for example, I exist, $7+3=10$, and it's better to be smart than stupid.[8] Anselm's *De casu diaboli* may offer a relevant example of claims based on intuition. Here Anselm assumes that, were an agent motivated by only a single, God-given motivation, he could not be praiseworthy or blameworthy. Why not? Because he cannot do otherwise. He wills by a responsibility-undermining necessity.[9] He and his student interlocutor canvas various particulars of the single-motivation hypothesis, but he does not attempt any proof or evidence that the agent in the proposed situation would not be free. Were he pressed, I suppose he would insist that he "just sees" that this is the case. And, given the importance of the assumption, and the consistency with which Anselm holds it, my suspicion is that he might allow that he "just sees" it with about the same clarity and certainty that Augustine attributes to the value claim, "It's better to be smart than stupid."

[7] I am sympathetic to Joel Pust's point (*Stanford Encyclopedia of Philosophy*, "Intuition") that principled skepticism about intuitions leads, by principled steps, to general skepticism. And skepticism is the wrong road for anyone, like Anselm, who takes it that the point of philosophy is to lead the good and happy life here and in the hereafter.

[8] *De libero arbitrio* Book 2.3, 8. For Augustine and Anselm our "just seeing" requires divine illumination. A standard puzzle for contemporary intuitionists is how—in the absence of any plausible connector between the knower and the known—human beings are capable of grasping necessary truth. See my "Evidence for God from Certainty." *Faith and Philosophy* 25 (2008): 31–46.

[9] DCD 13 and 14. In *De libertate arbitrii* (DLA) 8 he argues that God cannot (*logically* cannot) be the cause of sin. In that sin happens, it follows that the created agent himself must be the cause of sin. This brief argument supports the basic intuition.

Obviously, with intuitions, there are degrees of "seeming." I find that the intuition that I exist (if it is proper to call this introspective "just seeing" intuition) and that $7+3=10$ are of such a degree of certainty that nothing could undermine them. I share Anselm's intuition that the singly motivated agent should not be held morally responsible, and find it almost as unshakeable as $7+3=10$. Other "seemings" might be more easily undermined. Anselm, good analytic philosopher that he is, is given to hypothetical examples and thought experiments, where the intuitive response to the example or experiment is used as data for the case Anselm hopes to make.[10] These intuitive responses are obviously not possessed of the same sort of certainty as $7+3=10$. Still they are helpful in making the various points Anselm is trying to get across.

The Anselmian, then, with the imprimatur of Anselm's employing intuitions of various sorts, is quite happy to participate in the usual back and forth of competing intuitions that characterizes the contemporary free will debate. But I would like to propose a methodology relevant to the use of intuitions, and that is that participants in the debate be alert to the fact that some of their intuitions (not, I take it $7+3=10$, but those held somewhat less certainly and universally) may be significantly colored by their background worldview. Sometimes what may be proposed as a brute or basic intuition concerning free will and responsibility may actually be an entailment (or at least some sort of consequence, possibly not recognized as such) of an underlying and more general view of the way the world is. For example, Daniel Dennett famously asks, "Why Do We Want Free Will?"[11] His answer is roughly that we want to be able to punish those who offend. We need enough free will to ground the appropriateness of such punishment, and a theory which gives us that sort of freedom has given us all we should want.[12] But his answer is obviously and wildly unsatisfactory to many of "us," probably because some of "us" just don't see the world the way Dennett does. In asking why "we" want free will, it is useful to consider who "we" are and what "we" think about things in general.[13]

[10] For example, how would you analyze the situation if you saw a man, whom you knew to be strong enough to hold on to a bull, trying and failing to hold on to a sheep (DLA 7)?

[11] Daniel C. Dennett, *Elbow Room* (Cambridge, Massachusetts: MIT Press, 1984).

[12] Many contemporary philosophers take this approach. For example, R. E. Hobart, who provides the recent *locus classicus* for the luck problem against indeterminism ("Free Will as Involving Determination and Inconceivable Without It." *Mind* 43 (1934): 1–27, at 19ff). Interestingly, this compatibilist justification for society's punishment of the agent whose will is determined was around at least as long ago as the thirteenth century. It is condemned by Stephen Tempier the Bishop of Paris in the Condemnation of 1277, #165.

[13] There is some reason to suspect that folk intuitions in favor of indeterminism are colored by a belief in God. Adam Feltz, Edward T. Cokely, and Thomas Nadelhoffer, "Natural Compatibilism versus Natural Incompatibilism: Back to the Drawing Board." *Mind and Language* 24 (2009): 1–23, at 15.

Sometimes, what prima facie look to be intuitive disagreements about free will issues may actually be, or at least stem from, disagreements about something more fundamental. So, for example, suppose one subscribes to the basic Epicurean worldview (setting aside his infamous "swerve"). Then one takes the world and all it contains to be fundamentally atoms and the void. The human being is atoms briefly configured this-person-wise for the life of this person. One's goal in life is to maximize one's own pleasure and minimize one's own pain. (Not through self-indulgence, but through moderation and self-discipline, as Epicureans always hasten to add!) On Epicureanism "justice" is the agreement among members of the community not to harm others in return for not being harmed themselves. There is nothing intrinsically wrong with being "unjust" and failing to abide by the agreement. But you procure your pleasure more successfully by refraining from injustice because you might get caught and punished, and even if you don't get caught, you can never be sure that you won't be caught, and so you would live in fear. On this worldview, enough freedom to ground punishment may be as much freedom as one might want.

When the connection between the worldview and punishment is spelled out this way, the Anselmian agrees with Dennett. *If* this Epicurean picture, or some contemporary update, of the world and man's place in it were the correct picture of the way things are, then Dennett might be right about what ought to count as "enough" freedom. But if the world is not as Epicurus saw it, then perhaps we should want more, or different, freedom. Uncovering the connections between one's background worldview and one's "intuitions" on free will can bring clarity to the issues involved. I suggest, as a general methodological principle, that when confronted with dueling intuitions, philosophers ought to reflect upon whether or not they arise out of more basic commitments.[14] I do not say that noting the connection between the basic worldview and the free will beliefs and intuitions is likely to produce more consensus on free will and related topics. It may lead disagreeing philosophers to hold that they have even less in common than they might have supposed. But clear-sighted and mutually understood serious disagreement is probably a more worthy goal in philosophy than a fuzzy minor

[14] George Sher engages in a quick example in discussing whether or not blame is a matter of belief. He writes, "it is tempting to dismiss the view that each bad act stains the agent's soul or reduces his moral balance as a calcified remnant of a now largely discarded religious morality. [We should not do that because]...even falsehoods can be widely believed and even incoherencies embraced *as* beliefs." *In Praise of Blame* (Oxford: Oxford University Press, 2006):76. Sher writes from a perspective of rejecting traditional Western religion, and it would be interesting to have a lengthier discussion of how the tradition may have produced our intuitions about blame and how rejecting that tradition affects how we should assess blame. Surely Sher is right that blame looks very different in a theist, as opposed to an atheist, universe.

disagreement or even a fuzzy consensus. A running theme throughout this work will be how the Anselmian's background Christian theism may affect his intuitions regarding free will. The suggested effect can take different forms. Certainly it involves attitudes which follow plausibly from a clearly held and well-defined belief in a just, omnipotent, and omniscient God. But Christianity as a historical tradition also includes recurring narratives, about conversion and forgiveness, for example, which may well color how the Christian philosopher responds intuitively to the myriad stories and extensive cast of characters which philosophers present to motivate intuitions about free will and responsibility. I hope to bring some clarity to the battle of intuitions by suggesting possible links between the Christian worldview and the Anselmian intuitions.

Another reason for being mindful of Anselm's classical theism—the belief that there is an omniscient, omnipotent, perfectly good God sustaining everything in being from moment to moment—is that it provides a useful structure within which to discuss free will, whether or not one takes it to be the true picture of the world. For one thing, it enables us to make a clear distinction between two questions that are sometimes run together in the literature. The question, "Should we—your fellow members of society—hold you responsible?" is not quite the same question as, "*Are* you responsible?"[15] We suffer under severe epistemic limitations, and that significantly affects our ability to judge whether or not you are responsible. (More on this in Chapter Five.) The hypothesis of the ideal observer, such as God, can prove useful in distinguishing these two questions and answering them separately.

Furthermore, much of the contemporary free will debate swirls around examples involving evil, controlling neurosurgeons and other assorted mad manipulators. And then the discussion can be misdirected into minutia concerning the limitations of said neurosurgeons and manipulators, missing more important points. And intuitions may be colored by the wickedness, or sheer strangeness, of these characters. In many cases, replacing these peculiar constructs with a well-known God brings clarity to the arguments and sharpens intuitions. I defend this point further in Chapter One, in which I offer a controller argument that is similar to others in the contemporary literature, but considerably stronger by hypothesizing a *divine* controller.

While Anselm's theism will be operating as a useful background assumption, his dualism will play no role in this exposition of his libertarian system.[16] Outside

[15] Some philosophers simply deny this distinction. R. E. Hobart (1934: 25) writes, "A man deserves punishment or reward if society ought to give it to him."

[16] Anselm died before he was able to write a proposed treatise on the soul. Likely he was an Augustinian style of dualist; soul and body are separable substances, but the complete human being

of philosophy one often finds people who assume that libertarian freedom requires substance dualism.[17] There is currently no consensus among philosophers on what used to be called the mind/body problem. Consciousness apparently remains a "problem." I have my doubts about physicalism, but for the sake of argument suppose that it is possible to successfully analyze the conscious actions of agents, actions like making choices, as entirely physical phenomena. If so, there is nothing in my presentation of Anselmian libertarianism that would conflict with a physicalist account of human choice beyond the basic problems that consciousness per se presents.

Terms and Usage

I have called the theory I am propounding, "Anselmian libertarianism." The theory includes a rich set of criteria constituting, or required for, or closely related to, a free choice. Among the criteria are the required antecedents of a free choice and the subsequent effects of that choice, especially regarding self-creation. By way of introduction we can note two key elements. The first, and most important, is that the agent chooses "from himself." In Anselm's parlance, he chooses *a se* or with aseity. Chapter Three attempts to spell out how this works. The second key element is that, preceding the choice, the created agent confronts genuinely open options. This second element is an entailment of the claim that the free choice is *a se*. In that the created agent is not the author of his own existence and properties, in order to have anything of importance which can truly be "from himself," he must choose between open options. Some philosophers propose views which they refer to as "libertarian," in that choices are said to be "self-caused" because they are produced by events internal to the agent. But they allow that these events may render a unique choice inevitable and that these events may be the product of other things and events for which the agent bears no responsibility. The choice, in this case, is not ultimately up to the agent.[18] My use of the term "libertarian" will not include such "qualified" versions, but rather insists upon choices originating absolutely with the agent, a requirement which in turn will entail an absolute ability to choose otherwise.

is a combination of the two. So, unlike with platonic or Cartesian dualism, it is not the case that the *real* you is to be identified with the thinking/consciousness part of you.

[17] For a striking example see Daniel Wegner, *The Illusion of Conscious Will* (Cambridge, Massachusetts: MIT Press, 2002).

[18] Eleonore Stump, for example, describes Augustine's position as a "qualified libertarianism," while allowing that, for Augustine, the internal events which produce a choice are caused ultimately by factors which are outside of the agent. See Eleonore Stump, "Augustine on Free Will." In *The Cambridge Companion to Augustine*, edited by Eleonore Stump and Norman Kretzmann (Cambridge: Cambridge University Press, 2001): 124–47.

Anselmian libertarianism is a "parsimonious agent-causation" view. As an inheritor of Aristotle's position that it is substances that bring things about through their powers, Anselm is a believer in—perhaps it would be better to say, he simply assumes—substance causation.[19] In bringing about choices, it is the agent himself who is the cause (by means of various "powers" as we shall see). Happily, the Anselmian need not be embarrassed by, or try to bracket, Anselm's commitment to substance causation, since the theory is currently making a comeback.[20] Anselm's agent-causation is "parsimonious," as I will explain in Chapter Three, because Anselm insists—and here he is at odds with most contemporary versions of agent-causation—that we must not add any new sorts of causation to the world just in order to explain how an agent could choose with aseity. Obviously, causation, in this Anselmian context, must be more than a thin, Humean constant conjunction. The Anselmian, like any libertarian, I take it, holds that, one way or another, the agent or agent-related phenomena *bring the choice about*. However we analyze that cause, it must supply the "oomph" which produces the effect.[21] For the purposes of the present work, "causation" will refer to the "oomph" sort of phenomenon. Anselm's claim is just that there is no "oomph" which is unique to the process of a rational free choice. (I grant that use of the technical term "oomph" does not provide an analysis of what is meant by a cause. In Chapters Two and Three more will be said about Anselm's understanding of how the will "moves" such that a choice is brought about.)

The Anselmian insistence on aseity entails that if a choice is made necessary by anything except the agent himself, it is not a free choice. There are many species of necessity that conflict with freedom. The usual debate in the modern and contemporary literature is between freedom and determinism. By and large I will use "determined" to mean "causally necessitated," since that meaning is the most important for my purposes. Some event, y, is causally necessitated by x, if x is some acting thing or some event, such that, given x, the unique event y *must*

[19] Anselm does have a little to say about how we use causal terms, especially in his *Philosophical Fragments*, but his discussion is more about word usage than the metaphysics of causation.

[20] See for example, E. J. Lowe, *Personal Agency* (Oxford: Oxford University Press, 2008); Helen Steward, *A Metaphysics for Freedom* (Oxford: Oxford University Press, 2012).

[21] To my knowledge almost all of the free will debate assumes the "oomph" theory of causation. The concern has been whether or not *the productive source* of a choice is the agent, is events within the agent, or is ultimately outside of the agent. Alfred Mele has made some attempt to bring a Humean analysis into play; *Free Will and Luck* (Oxford: Oxford University Press, 2006): 194. See also Helen Beebee and Alfred Mele "Humean Compatibilism." *Mind* 111 (2002): 201–24. For a discussion of some moral difficulties entailed by a Humean theory of human choices and actions see the fourth section of my "What's Wrong with Occasionalism?" *American Catholic Philosophical Quarterly* 75(2001): 345–69.

happen, because x *makes* it happen.[22] So say that Bill commits adultery, and he does so because God produces in Bill an irresistible desire. Several, mutually consistent, locutions involving "causal necessity" would be apt in describing the situation. I might say that "God, by producing Bill's irresistible desire, causally necessitated Bill's adultery." I might say that "Bill's having the irresistible desire causally necessitated his act of adultery." I might say that "Bill's succumbing to his irresistible desire causally necessitated his act of adultery." Given God's producing the desire and Bill's having it and his inevitably succumbing (it's an irresistible desire), the act must happen. It is *made* to happen by God's producing and Bill's having and succumbing. The locution that I would avoid in this case is, "Bill causally necessitated his act of adultery due to his irresistible desire." The reason this is not the right description of the event is that, on Anselmian libertarianism, if Bill couldn't help but commit adultery because of his God-given, irresistible desire, then it wasn't really Bill himself who made it happen.

Often "determinism" has been taken to mean, "universal determinism," the thesis that "a complete statement of the laws of nature together with a complete description of the condition of the entire universe at any point in time logically entails a complete description of the condition of the entire universe at any other point in time."[23] This situation might hold on various theories of causation, but when I use the term "universal determinism" I mean that the complete description at one time entails a complete description at another *because* every event is causally necessitated. Universal determinism (except for universal *theist* determinism, the thesis that God causes everything that exists and everything that happens) was not an issue in Anselm's day, and is no longer the consensus among scientists and philosophers. But even if it is not the case that *everything* is determined, it might still be the case that human choices are all determined, and that is the concern here. "Natural determinism" will mean that, even if it is not the case that every event is causally necessitated by the past and the laws of nature, the relevant events, human choices, are causally necessitated by the past and the laws of nature. But even if natural determinism is false, it could nonetheless be the case that human choices are causally necessitated by something other than the past and the laws of nature.

By a choice being "determined," like adulterous Bill's choice in the example just given, I will mean that the choice is "causally necessitated by factors not ultimately originating from, or identified with, *a se* choices of the

[22] I do not attempt to develop or elaborate on the very difficult metaphysical question of exactly what it is to *make* something happen.

[23] Alfred Mele, *Effective Intentions* (Oxford: Oxford University Press, 2009): 150.

conscious agent."[24] This means that there can be indeterminism in the history of a choice, and yet the choice be determined. For example, suppose a certain choice is the causally necessitated product of the *non*-determined behavior of some subatomic particle, which behavior does not ultimately originate from, or cannot be identified with, an *a se* choice of the agent. That choice is determined, by my definition. (I do not rule out the possibility that, on some versions of physicalism, the behavior of a particle might *be* the *a se* choice of the agent, in which case the choice is not determined. Nor would it be correct, on this possibility, to say that the behavior of the particle "causally necessitated" the choice, unless it is allowable to say that some event may causally necessitate itself, which seems a peculiar and unhelpful locution.)

That a choice is determined does not necessarily mean that it is causally necessitated by something in its past. If a mad neurosurgeon from the future sends a signal back in time that necessitates a present choice of yours, then your choice is determined. And if a choice is causally necessitated immediately by God—God brings it about that you choose A at the time you choose A, such that you cannot choose otherwise—your choice is determined. And if a choice is causally necessitated by a motive (or desire or intention), that is, having this motive makes it inevitable that you make this choice, where the ultimate source of the agent's having the motive is not the agent, then the choice is determined.

There is another sort of necessity, besides *causal* necessity, that conflicts with *a se* choice. On the Anselmian account the only possible grounding for the truth about, or knowledge of, which option in an *a se* choice is actually chosen, is the *actual* agent's *actually* making the *actual* choice. Period. I will call this claim, the "grounding principle." I spell it out in Chapter Four, and it will bear significant fruit in addressing standard problems in the free will debate. There are theories—even nominally libertarian theories—that include the claim that the universe is such that there is a truth about, or knowledge of, a free choice, where this truth or knowledge is not grounded in, and dependent upon, the making of the actual choice itself. But the Anselmian insists that such theories introduce a necessity which conflicts with aseity. These theories entail that it is "externally non-causally necessary" that the agent choose one way rather than another. I will label this sort of necessity "ENC necessity" for short.[25] On such a theory, even if the universe is

[24] This is a narrow definition to serve my purposes in this work. I do not intend to imply that an agent-causal choice which does not meet the exact formulation of an Anselmian *a se* choice must ipso facto be determined.

[25] The label is awkward, but no preferable alternative occurs to me. One might suggest, "metaphysical" necessity, but in some circles that has come to mean "true in all possible worlds," and that is not what I mean.

such that your choice for this over that is not causally necessitated, your choice is necessitated, and not by you, the agent. So, for example, if there exist true propositions about what you will choose in the future, and the future is at present absolutely non-existent, so it is not your actually making the choice that grounds the truth about your future choice, then you will not be making that choice with aseity. It will not be a free choice as the Anselmian understands free choice, since the universe is such that you "must" choose as the preexisting truth entails that you will choose. (In Chapter Four I briefly note Anselm's solution to the dilemma of freedom and divine foreknowledge; all times exist and it *is* your actually making a future free choice which grounds God's knowledge of your choice.) If you choose with aseity your choice can be neither causally nor ENC necessitated.

There are at least two species of necessity regarding choice which do not undermine the aseity of the choice, where both involve the necessity originating with the agent.[26] There is what Anselm refers to as a consequent (*consequens*) necessity. This is the logical necessity that follows from the positing of some event. If X, then, as a matter of logical necessity, it follows that X. Suppose you are reading right now. Then it follows that you are reading right now, and it is impossible that you not be reading right now. But this necessity does not conflict with aseity. If you made an *a se* choice to read, then the fact that it is consequently necessary that you be reading when you are reading is dependent upon your choice. Or suppose, as suggested above, that the presently existing truth about what you will choose tomorrow is actually, today, grounded in, or dependent upon, what you choose *a se* tomorrow. Then *you*, by choosing, have made it consequently necessary today that you choose as you do tomorrow. But this does not undermine your aseity, since it was up to you to originate the consequent necessity.[27] So, whereas an ENC necessitated choice cannot be *a se*, a consequently necessitated choice can be. This important distinction will help solve some problems later in this work.

A second sort of necessity which is consistent with aseity can be called "self-determinism" or "character-determinism." Remember that the point of free will, for Anselm, is that we should be able to contribute to our own "creation." Anselm holds that a choice which is determined by the agent's character may be considered *a se* if the character itself is the product of the agent's past *a se* choices. Perhaps, as you make your present choice, your

[26] I avoid saying that the agent, who chooses with aseity, causally necessitates his choice by choosing, because that could suggest that there is an act of choosing which precedes the *a se* choice and then we open the door to a regress. The agent is the absolute cause of his choice, but not by some preceding act or choice.

[27] Rogers (2008): 170–6.

character renders it inevitable that you choose A rather than B. Nonetheless, Anselm and the Anselmian hold that you choose A with aseity, and are responsible for the choice, if it was ultimately up to you that you have the character that determines the choice. The thought that you are responsible for the actions which grow from your character, in that you formed your character through your actions, goes back at least to Plato and Aristotle, is shared by many contemporary libertarians, and is ineliminable from Anselm's overall theory of freedom. Recently difficulties have been raised with the position, and these will be addressed in Chapter Nine.

To offer some remaining definitions: "compatibilism" will mean the position that a human choice can be free in some important, responsibility-grounding, way even if it is determined and/or ENC necessitated. (I take it that, insofar as choices are concerned, many versions of determinism entail ENC necessitation as well.) Note that this is a somewhat more expansive definition than usual in the literature, in that it covers ENC necessity. "Incompatibilism" is the view that compatibilism is false. Both libertarians and "hard" determinists are incompatibilists. "Hard determinism" takes it that choices are determined and/or ENC necessitated, compatibilism is false, and so no human choices can be free in a very robust, responsibility-grounding way.

Evidence from Science and Experience?

I entitled Part I of this Introduction, "Why Do We Want Free Will?" Anselm, as noted at the beginning of this Introduction, wants created agents to be free, since otherwise they cannot be *imagines dei*. I elaborate on this response below, but first it will be well to explain why it is legitimate to allow the question about what we *want* to drive the discussion about what we *have*. Someone new to philosophy and the free will debate might suppose that our first question ought to concern the *evidence*—preferably scientific—for whether or not human beings have free will, and, if we do have free will, what sort it is. The Anselmian insists that we have sterling reason to believe that we have free will—a robust (if parsimonious) libertarian free will!—but the reason is not based on evidence from science or from our ordinary experience of choosing. In defense of the Anselmian approach, I will briefly outline the reasons for believing that neither science nor our introspective experience of choosing can decide the question between the libertarian, the compatibilist, and the "hard" determinist.

A hundred years ago many intellectuals took universal determinism to be the scientific approach. Nowadays the consensus is that the universe is shot through with indeterminism in that sub-atomic particles behave indeterminately. This micro-indeterminism is not relevant to our descriptions and understanding of simple macroscopic systems, such as the solar system. We can, for example, predict solar eclipses indefinitely into the future, with extreme accuracy, even though all those constitutive sub-atomic particles are behaving indeterminately. However, in some systems the micro-indeterminacy can get amplified into macro-indeterminacy. For example, the indeterminate decay of a radioactive nucleus causes an indeterminacy in when a Geiger counter emits a click. Some "chaotic" systems, such as the weather, whose behavior is very sensitive to very slight changes—the "butterfly effect"—may also allow such amplification. Even an ideal knower who knew all the physical facts at the moment and all the laws of nature—even God, if He were in time and had to extrapolate the future from the present—could not know what the weather will be far in the future.[28] Intuitively it seems likely that human behavior, being very complex, is more similar to the weather than it is to the solar system, but that alone is not enough to rule out the possibility that on the macro-level of human behavior our choices are determined. Contemporary physics does not decide the question.

And neuroscience, while making huge strides, is still in its infancy. It cannot deliver an empirical conclusion on the question of what, if any, free will we enjoy.[29] Indeed, the thought that what we need to do is, "develop an understanding of 'freedom' that is consistent with contemporary neuroscientific understanding of the brain and behavior"[30] seems exceedingly premature, depending on what one meant by "consistent with." If "consistent with" means only that neuroscience has not demonstrated much about choices, including whether or not they are determined, then fine. Any view of free will is "consistent with" contemporary neuroscience. If "consistent with" means that one ought to adopt a view of free will that allows one to *explain* free choice—what it is and how it happens—within current neuroscientific theories, then that seems far too limiting. Take this analogy: Suppose I hold that 7 + 3 = 10 is a necessary truth, and suppose I claim to know indubitably that, necessarily 7 + 3 = 10. I am reasonably sure that contemporary neuroscience does not propose a developed and satisfactory explanation of how I could indubitably know the necessary truth that 7 + 3 = 10. Should I then abandon my claim? I'm betting that my claim about 7 + 3 is, and ought to be, a far more secure epistemic possession of mine than *any*

[28] I believe there may be debate about this claim, but my local particle physicist, Stephen Barr of the Physics Department at the University of Delaware, takes this to be the current consensus among physicists.

[29] Colin Klein, "Philosophical Issues in Neuroimaging." *Philosophy Compass* 5 (2010): 186–98.

[30] William T. Newsome "Human Freedom and 'Emergence'." In *Downward Causation and the Neurobiology of Free Will* edited by Nancey Murphy, George F. R. Ellis, and Timothy O'Connor (Berlin: Springer-Verlag, 2009): 53–62, at 53.

part of *any* neuroscientist's cognitive theories. In fact, I'm willing to say that if I must doubt my belief that necessarily $7+3=10$, since neuroscience falls short of explaining it, then no claim to knowledge is secure and science is untrustworthy. Our epistemology should not be hostage to the limitations of contemporary neuroscience. But why should metaphysics suffer under a circumscription from which we exempt epistemology? Given the long history of metaphysics, and the youth of neuroscience, it is unreasonable to allow the latter to dictate to the former.

There are some outside the philosophical community—some experimental psychologists and media folk reporting on the findings of these experimental psychologists—who claim that experimental psychology has shown, or is showing, that human beings do not have free will.[31] The philosopher engaging with this literature is dismayed to discover that these psychologists are unfamiliar with philosophical definitions or descriptions of free will. Indeed, they do not even offer their own definitions. While the experiments in question are often interesting and helpful regarding the workings of the human mind, they do not even connect with the philosophically sophisticated free will debate. Much less do they justify the conclusion that free will, under any plausible definition, does not exist. (Perhaps the silver lining here, from the philosopher's perspective, is that philosophy obviously has a great deal to contribute to experimental psychology in the critical thinking area.) Alfred Mele has devoted a book to pointing out this gap between the evidence and the anti-free will conclusions.[32] Mele approaches the question from a compatibilist perspective, but it is safe to say that the evidence does not count against libertarian free choice, either.

A defense of this claim lies beyond the scope of the present work, but it should be noted that demonstrating experimentally that no agent could possibly choose other than they do choose would be a daunting task. Many libertarians—this is certainly true in the Anselmian account—are quite willing to admit that the free choices they envision may occur only rarely. So it would take a very significant number of correct predictions, in the sorts of cases where the agent is making what look to be free choices under some sophisticated definition, in order to plausibly generalize that no one could ever choose other than they choose. It is

[31] The work of Benjamin Libet is often cited as evidence that choices are determined. Libet's experiments, while informative, do not provide evidence for determinism. And Libet himself never claimed they did. See his "Do We Have Free Will?" for an overview of his work (*The Oxford Handbook of Free Will* edited by Robert Kane, First Edition (Oxford: Oxford University Press, 2002): 551–64).

[32] Mele, (2009). See also Eddy Nahmias, "Scientific Challenges to Free Will." In *A Companion to the Philosophy of Action*, edited by Timothy O'Connor and Constantine Sandis (Chichester, UK: Wiley-Blackwell, 2010): 345–56.

safe to say that for now, the foreseeable future, and perhaps forever, science leaves open the question of whether or not libertarian free will exists.[33]

Some experimental psychologists have offered what might be considered a sort of pragmatic argument for determinism. It goes roughly like this:

1. We are successfully engaged in the science of human behavior.
2. If human behavior, including human choices, is not determined, then we cannot successfully engage in the science of human behavior.

Therefore human behavior, including human choices, is determined. Q.E.D. [34]

Of course, someone who accepted Premise 2 could comfortably deny Premise 1. But, happily for the progress of sciences of human behavior, Premise 2 is obviously false. As I noted above, many of the experiments which experimental psychologists perform provide very useful information about how we human beings exercise our power to will, think about choosing, and go about making choices. But, just as none of these experiments *proves* determinism, none presupposes that determinism be true, or believed to be true, as a condition of conducting the *actual* experiments.[35] Perhaps, in that psychologists may be hoping to find causal explanations for behavior, a sort of working assumption that there are causes to be found is beneficial and appropriate for the experimental psychologist.[36] But the psychologist would do well to be clear that a helpful working hypothesis does not equate to a true metaphysical proposition.

And scientists should be cautious. We all know that Science (I capitalize to indicate the popularized version one finds in the news media), in the service of unwholesome ideologies, has inflicted great harm on humanity in the recent past. One can easily imagine some politician asking himself rhetorically, "Well, now that I've heard that nobody has free will, so nobody is really in control of themselves anyway, wouldn't it be better if I take control over people's lives, since I know best what's good for them?" In the mouth of the scientist the

[33] Mark Balaguer seems sanguine that the future does indeed hold a scientific proof one way or the other on the question of whether or not choices are actually determined. He writes, "the libertarian question reduces to a straightforward empirical question about the physical world." And assumes that this question will be answered. To my mind he does not offer a plausible defense for his optimism. *Free Will as an Open Scientific Problem* (Cambridge, Mass: MIT Press, 2010): 21.

[34] See for example, George S. Howard, "Whose Will? How Free?" In *Are We Free? Psychology and Free Will* edited by John Baer, James C. Kaufman, and Roy F. Baumeister (Oxford: Oxford University Press, 2008): 260–74, at 261; See also John Baer, "Free Will Requires Determinism." In Baer (2008): 304–10, at 309.

[35] See Roy Baumeister's four reasons for refusing to be bullied into embracing determinism, "Free Will, Consciousness, and Cultural Animals." In Baer (2008): 65–85, at 67.

[36] Shaun Nichols, "How Can Psychology Contribute to the Free Will Debate?" In Baer (2008): 10–31, at 22.

unjustified denial of free will may inspire who knows what bad behavior on the part of those who are itching to take control of their fellows.[37] If libertarianism is true, then psychology will never be able to explain, much less predict, each and all of our choices. That need not spell the end of psychology. Psychologists may just have to live with less exactitude than they might like. But, as Aristotle says, it is unhelpful to insist on more exactitude from a subject than the subject allows.

Sometimes philosophers, looking for evidence for free will, appeal to our conscious experience of debating between options and then choosing. Don't we recognize in an ineradicable way that we could opt for this *or* that, and isn't that immediate evidence that our choices are not determined? After we've chosen, unless we are in the grip of a philosophical theory, aren't we convinced that we "could have chosen otherwise"? To my knowledge no one disputes these phenomenological phenomena. Were libertarianism not thought to be subject to serious or insurmountable philosophical difficulties, this experience might be widely held to be good prima facie evidence that libertarianism is true. But there are those purported difficulties, and our experience of deliberating and choosing is quite compatible with the thesis that all of our choices are determined. We may *feel* that we could have chosen otherwise, but perhaps that feeling is just mistaken.[38]

A different sort of introspective evidence is worth mentioning here. In asking why we want free will, I will be making the traditional argument, pioneered by Anselm, that free will is necessary—a very robust, libertarian free will!—if we are to be the special, responsible sorts of beings that we usually hold ourselves to be. But why should we "hold" ourselves to be special and responsible? Well, don't we often believe we have done wrong? We are ashamed. And not in some, "gee-it's-too-bad-that-happened" way, but in a non-negotiable "I-shouldn't-have-done-that!" way.[39] And aren't we angry when we are falsely

[37] I document reasons for concern in "Freedom, Science, and Religion." In *Scientific Approaches to the Philosophy of Religion* edited by Yujin Nagasawa and Erik J. Wielenberg (London: Palgrave Macmillan, 2012): 237–54.

[38] Recently Helen Steward (2012) has suggested that agency, whether human or animal, requires the agent to be able to act in such a way as to settle an open future. She argues that there do seem to be agents—indeed we have evolved to be able to distinguish agents from non-agents—so determinism is likely false. I am entirely sympathetic to Steward's conclusions. It seems to fly in the face of ordinary experience and good sense to deny agency in which agents act indeterminately to settle an open future. But this is exactly what compatibilists and hard determinists do deny, and I do not see anything in Steward's argument which is likely to budge them.

[39] Derk Pereboom argues that the it's-too-bad-that-happened sentiment or belief is enough to ground and explain our attitudes; *Living Without Free Will* (Cambridge: Cambridge University Press, 2001): 204–6. I think this is false, but what I am claiming here is that what we actually *see* in ourselves is more than the unhappy occurrence of the universe unfolding in a regrettable way.

accused or when our good deeds are ignored or undervalued? Perhaps I do not "see" that I am a morally responsible being with the same indubitable clarity that I see that 7 + 3 = 10. But I see it pretty clearly! And I find that I recognize in others what I see in myself. Of course, the determinist and compatibilist are free to opine that what I think I "see" is really an illusion. But in that the scientific *evidence* does not justify any presumption in favor of determinism, the libertarian may justifiably resent being told that she really doesn't see what she believes she sees. The libertarian may counter that the recently developed event-causal and agent-causal libertarian views, even with all of their difficulties, are one and all more plausible than the claim that I do not have the sort of responsibility I see myself to have. And if a more successful libertarianism can be presented, as I hope to do in this work, so much the better! Those who claim that responsibility-grounding free will is really an illusion often add that it is an illusion that we need and that we cannot root out of our thinking in any case.[40] But if the belief is so important and ineradicable, perhaps that is reason to consider (or reconsider) the possibility that it is true.

In the contemporary literature the defense of hard determinism or compatibilism usually begins with an attack on libertarianism. It is argued that, since libertarianism is incoherent or unsuccessful in securing its aims, hard determinism, or, more popularly, compatibilism, must be true. And then the project is to try to save what needs saving out of the human condition on the hypothesis that all of our choices are, or may well be, determined. I said above that the Anselmian holds that we have sterling reason to embrace libertarianism. The reason is that, if we are actually the sorts of things we see ourselves to be, and certainly if we are the *imagines dei* which Anselm, for theological reasons, takes us to be, then we must be free in a libertarian sense. So we had better not abandon the project of constructing a coherent and successful libertarianism. Another way to put this is that we *cannot* save what needs saving out of the human condition on the hypothesis that all of our choices are determined. Which brings us back to the

[40] Daniel Wegner, for example (2002), argues that human actions are determined and conscious will is epiphenomenal. He justifies his belief in determinism (p. 54) by appealing to experiments by Benjamin Libet, although Libet himself does not believe that his experiments provide evidence for determinism (see Libet (2002)). Wegner also discusses a plethora of experiments in which people make assorted mistakes about what it is they are doing. Wegner concludes with the thought that conscious will is an important illusion that is not going away (p. 342). Wegner repeats this point in "Self Is Magic." In Baer (2008) 226–47, at 238–44. Saul Smilansky offers a more philosophically sophisticated defense of the claim that free will is an illusion; *Free Will as Illusion* (Oxford: Oxford University Press, 2000). But he, too, allows that it is an illusion that we need; "Free Will: Some Bad News." In *Action, Ethics, and Responsibility* edited by Joseph Keim Campbell, Michael O'Rourke, and Harry S. Silverstein (Cambridge, MA: A Bradford Book, MIT Press, 2010): 187–201.

question, "Why do we want free will?" or more accurately, "Why do we want the sort of very robust free will that requires aseity?"

Freedom and Dignity

Contemporary work on free will is often introduced with the thought that, as we theorize concerning the human condition, we require that you have free will in order that we be justified in punishing or rewarding you for some overt act you have done. (Throughout this work I will consider making a choice to be an act, such that the physical expression of that choice can be called the "overt" act.) Maybe (as with Dennett mentioned in this Introduction in the section titled "Worldviews and intuitions") this requirement has to do with how punishment and reward will benefit you and others. Maybe it has to do with whether or not you *deserve* to be punished or rewarded. Or perhaps we require that you have free will in order to justify our praising or blaming you for some overt act you have done. Or, phrased slightly differently, in order to justify our holding you morally responsible for your deeds. Or perhaps we require that you have free will to explain how our reactive attitudes, gratitude, indignation, forgiveness, etc., can be properly focused.

From the Anselmian perspective it is true that your having free will is necessary in order for you to be deserving of punishment or reward, of blame or praise, of our holding you morally responsible, and of our adopting the reactive attitudes towards you appropriately. But none of this gets to the heart of the matter or answers the question, "Why do we want free will?" Here I offer a brief look at how Anselm himself answers the question from within his worldview of classical Christian theism?[41] Some contemporary philosophers will embrace his entire approach wholeheartedly. Some contemporary philosophers will reject some aspects, but find others appealing and plausible. Some contemporary philosophers will find this way of going about things wildly wrong-headed.[42] But it

[41] Robert Kane in *The Significance of Free Will* (Oxford: Oxford University Press, 1996) offers a list of ten standard reasons for wanting a robust free will (pp. 81–9). The Anselmian can endorse this list, with certain qualifications, but it does not include the point about metaphysical value, which the Anselmian takes to be fundamental. It mentions "dignity" but the understanding of dignity is somewhat different from Anselm's. Kane associates dignity with Kantian autonomy, being a self-legislator. Anselm and his followers would not embrace such a concept.

[42] Steward (2012) would fall into this category, I believe. She faults those who begin the free will discussion from the perspective of human value and morality and responsibility. The problem with beginning here is that the concept of free will that will be required to make sense of these issues will be far richer than the agency-indeterminism she ascribes to animal agents. In response to this criticism I fear I must plead guilty. Or *very* guilty.

is important to explain the motivating foundation of Anselmian libertarianism in order to better understand the theory that is built upon it.

The "human condition" about which contemporary philosophers discussing free will theorize has been significantly influenced by Christianity, not just in the Western world, but anywhere Western culture has sunk in even a little, which means everywhere on earth, with the possible exception of those protected islands where everything from the outside is forbidden, and the inhabitants shoot arrows at passing planes. In the West many of our ethical principles, our social mores, and our legal and political institutions, have grown out of a Christian culture. (A prime example is the claim, derivable from many New Testament sayings, and entrenched in Western thought by St. Augustine, that all human beings are ultimately equal and of great, inherent value.)[43] One part of the basic project of the contemporary free will debate consists in attempting to offer theoretical support for these principles, mores, and institutions, and the common intuitions associated with them. Many of our contemporaries, philosophers and folk alike, may well believe what they believe about free will, and have the intuitions they have, in part because they share some or all of Anselm's Christian assumptions. And many who do not share any of those assumptions nevertheless embrace some or all of the principles and institutions which grew out of that Christian culture. And some who wholly reject the principles and institutions do so, in part, *because* they reject the Western and Christian culture which nurtured them. Thus, exposing the Christian roots of Anselmian libertarianism can help to bring clarity to the free will debate, and may provide at least one reason why intuitions differ.

Anselm's version of "Why do we want free will?" is "Why did God give us free will?" That is, what *good* is it for us to have free will? His answer is that human beings are a remarkably important and special kind of thing. On his metaphysics, anything that exists at all has value, it is good just insofar as it exists. This was the standard medieval position, and is arguably the way scientists look at the world when actually engaged in their disciplines. Ask a particle physicist about the equations describing the behavior of sub-atomic particles, or a zoologist about the habits of naked African mole rats, or an astronomer about the non-local universe accessed through the Hubble telescope, and chances are he will wax poetic in his enthusiasm about the beautiful equations, or the insect-like rats, or

[43] In defense of this claim see my "Equal Before God: Augustine on the Nature and Role of Women." In *Nova Doctrina Vetusque* edited by Douglas Kries and Catherine Brown Tkacz, American University Studies, Series VII, Theology and Religion, Vol. 207 (New York: Peter Lang, 1999): 169–85.

the distant planets. Once you start paying attention, things are just interesting, exciting, and worthy of our concern. Things are good.

All things, insofar as they exist, have metaphysical value. And human beings have enormous metaphysical value in that we are especially reflective of the nature of God (much more than the particles, planets, and rats) because we have the splendid properties of being rational and free. The view that all human beings are tremendously, objectively, valuable and equally valuable—even the poor or ill or old, even the foreigner, even women and children—has (at least until recently) been a fairly widespread assumption in the Western world. It has been common to hold that each human being has a special dignity—a view that can be expressed in "human rights" language—and that an aspect of that dignity derives from the human capacity for autonomy. It is not just our rationality, but our free will, that makes us beings worthy of a kind of respect that goes far beyond the sort of concern we might properly direct towards lower animals.

St. Augustine, one of the chief architects of Western thought, makes this case firmly and vividly, and most of the great Christian philosophers since his day have followed his lead. Western principles and institutions have grown up within the assumption that human beings are free and that our freedom is a great-making property. We have in-the-eyes-of-God freedom, which entails moral responsibility. We can *deserve* praise and blame, reward and punishment, in some very strong sense of "deserve" that goes beyond its being merely useful to praise and blame, reward and punish us. We are fit subjects of reactive attitudes, even bracketing the well-functioning of society. We have a value which is *utterly* unlike that of a nice painting or a cool car.[44] I take it that this is the view that has been assumed by many intellectuals and most of the folk in the Western world for the last two thousand years. The traditionally minded among us want free will, not because it is a requirement of our system of punishment, but because it is a key ingredient in our unique human dignity.[45]

[44] Derk Pereboom, in an effort to save human worth in the absence of free will, suggests that "moral worth is indeed moral, but it is more similar to the value we might assign to an automobile or a work of art" (2001: 153).

[45] Humans before the age of rationality and the permanently cognitively impaired have traditionally been held to possess human dignity as members of the species. Anselm, as we will see, argues that freedom is necessary for the agent to be able to help in his own creation. Could it be that the greatest tragedy of the death of a young child is that he will never be able to participate in this process? Anselm himself insists upon a real unity of the human family. (See my "Christ Our Brother: Family Unity in Anselm's Theory of Atonement." *American Catholic Philosophical Quarterly* 86 (2012): 223–36.) Perhaps the young child can in some way participate through that unity. This raises interesting issues concerning baptism and grace. A meditation on this question lies outside the scope of the present work, but it would be a fruitful topic.

But what sort of free will provides the appropriate ground for human dignity? There is some evidence that, pre-reflectively, most of the folk and many intellectuals assume that a determined choice is not sufficiently free to allow for responsibility and the rest, so we may take it that they have libertarian leanings.[46] (One sort of evidence for this comes from those experimental psychologists mentioned in the section on "Evidence from science and experience?" who assume that if they can show that a choice is causally necessitated—not that they have done so!—they would be showing that people do not have free will.) But on the other hand many of the great Christian philosophers have been what I call "theist compatibilists." They hold that human beings make free choices for which they are responsible and justly rewarded or punished by God, and yet that these acts of choice are caused by God. So, for example, although on the one hand it seems appropriate to cite St. Augustine, in his *On Free Will*, as one of the great defenders of the importance of free will in Western thought, in the final analysis, at least later in his life, he grants that all our choices are ultimately caused by God.[47] Thomas Aquinas, too, emphasizes the importance of human free will, and yet agrees that God causes all our choices.[48] Prima facie, given the importance of free will in Christian thought, and allowing the suggestion that intuition points to a conflict between freedom and determinism, it seems surprising that such philosophical giants of Christian philosophy should adopt theist compatibilism. Augustine and Aquinas take this tack for two reasons. First, they hold it to be required by the view that God is absolutely omnipotent and sovereign. If little created agents exercise some sort of causality in independence of what God wills

[46] Nichols (2008): 13–14.

[47] See my *Anselm on Freedom* (2008): 30–52.

[48] I would consider Thomas a compatibilist, in that he holds that, while our choices are caused by something other than ourselves—God as primary cause—we are still responsible. Of course, our choices are also caused by us, as secondary causes. See *Summa Theologiae* 1, Q.83, art.1, ad. 3 and *Summa Contra Gentiles* 1:68. Brian Shanley, O.P. argues that, since the divine causation in question is not the temporally antecedent natural causation posited by contemporary determinism it is better not to label Thomas a compatibilist ("Beyond Libertarianism and Compatibilism: Thomas Aquinas on Created Freedom." In *Freedom and the Human Person*, edited by Richard Velkley (Washington, DC: Catholic University of America Press, 2007): 70–89). Hugh McCann has recently proposed a view very close to Thomas's. See chapters five and six of his *Creation and the Sovereignty of God* (Bloomington and Indianapolis: Indiana University Press, 2012). McCann has also developed this position in a series of articles: "Divine Sovereignty and the Freedom of the Will." *Faith and Philosophy* 12 (1995): 582–9; "Sovereignty and Freedom: A Reply to Rowe." *Faith and Philosophy* 18 (2001), 110–16; "The Author of Sin?" *Faith and Philosophy* 22 (2005): 144–59; "God, Sin, and Rogers on Anselm: A Reply." *Faith and Philosophy* 26 (2009): 420–31. I have responded in "Does God Cause Sin? Anselm of Canterbury versus Jonathan Edwards on Human Freedom and Divine Sovereignty." *Faith and Philosophy* 20 (2003): 371–8; "God Is not the Author of Sin." *Faith and Philosophy* 24 (2007): 300–10; "Anselm against McCann on God and Sin: Further Discussion." *Faith and Philosophy* 28 (2011): 397–415.

and causes them to do, then, argue Augustine and Aquinas, God is not wholly and absolutely in charge. Moreover, they subscribe to classical theism which insists that anything that can properly be called a "thing," anything with any ontological status at all, is caused to be and sustained in existence by God. And so, assuming an act of choice to be a sort of thing, it, too, must be caused to exist by God. If Augustine and Aquinas ever had the anti-compatibilist intuition, it was overridden by their commitment to the thought that God causes everything.

But if everything you do is caused by God, it seems unjust (as I discuss in Chapter One) that He (or anyone) should hold *you* responsible for what *He* caused you to do. Thus Anselm proposes a different view of free will. He embraces the assumption that a determined choice cannot be free *enough* to give us what "we" want out of free will. Regarding our metaphysical value, he moves us up the Great Scale of Being one more notch, beyond the, still pretty elevated, station at which Augustine's and Aquinas' theist compatibilism would place us. Anselm's claim is that, beyond simply being rational and free in some "modest" way, we (and the angels, about whom more later) are uniquely in the image of God in that, unlike everything else, we can participate in our own creation by making ourselves better *on our own*.[49] Augustine had explicitly dismissed this suggestion as absurd. But no, says Anselm, we have the (remarkable!) sort of free will which enables us to choose from ourselves, we have aseity. One aspect of God's perfection is that He exists absolutely independently, *a se*. Anselm holds that God has made us such that, through free choice, we can reflect—albeit in a very dim and limited way—divine aseity. (Anselm very clearly adopts the free will defense in response to the problem of moral evil. Why does God permit wicked choices? Because He has to permit morally significant choices in order to allow us to have the elevated metaphysical stature of self-creators, and we sometimes choose badly. But this is an issue that lies outside of the scope of the present work.)[50]

In the contemporary literature critics of libertarianism have sometimes said that it proposes an impossible demand, that the free agent should be some sort of creator *ex nihilo*. As I set out Anselmian libertarianism, it will become clear that the scope Anselm allows for created aseity is actually very limited. We did not

[49] *Cur deus homo* 2.10; see Rogers (2008): 56–9.

[50] See Chapter Four under "The Grounding Principle" for Anselm's rejection of Molinism. Anselm clearly does—literally!—accept "*heaven-and-hell* responsibility." The term is from Galen Strawson ("The Bounds of Freedom." In *The Oxford Handbook of Free Will*, First Edition, edited by Robert Kane, (Oxford: Oxford University Press, 2002) 441–60, at 451). Strawson finds the view morally repugnant. The Anselmian belief is that the worst torment of hell is being divided from God. And God leaves you free to reject Him (*Catechism of the Catholic Church*, section 1033) because freedom is such a great good.

bring ourselves into being, and we are mostly made by factors outside ourselves. And yet, in however circumscribed a way, we are unique and remarkable in that we possess aseity. For Anselm the whole point of our being able to choose freely is that it bestows upon us the objective value of being the sorts of things that can imitate God by contributing to the creation of ourselves. The reason why at least some of us want free will is this: We hold ourselves and other human beings to be morally responsible agents with an ability to engage in self-creation which gives us a unique and special value. And we recognize that this assessment of ourselves and others requires a robust free will. Contrary to some, if science should (somehow) prove that human choices are all determined, I would judge that no human agents are free and responsible enough to ground the elevated status traditionally ascribed to us. I surmise that Anselm would agree. In Chapter One I defend this position, in Chapters Two, Three, and Four, I spell out Anselmian libertarianism, and in subsequent chapters I respond to the criticisms that must be addressed if Anselmian libertarianism is to be considered a viable option.

Further Methodological Principles

I have already noted that I intend to alert the reader when it seems to me that Anselmian intuitions may be rooted in Christian assumptions, or in beliefs, attitudes, etc., which bear the imprint of Christian culture. Another principle already mentioned, but perhaps deserving a few more words here, is the point that the locus of our interest will be choices, not overt deeds. It is important to distinguish carefully between the agent's making a choice on the one hand, and the agent's engaging in the overt action that follows from that choice. This is a distinction that has been emphasized in the recent free will literature.[51] And yet it is still sometimes blurred, so it will be useful to spell it out and to see how it will function in the project of developing Anselmian libertarianism.[52] On the Anselmian account, an agent can properly be held morally responsible for his overt deeds only if they are the result of choices for which he is responsible, that means *a se* choices or choices determined by his character, when the character was responsibly formed through *a se* choices.

[51] For example David Widerker, in responding to Frankfurt-style counterexamples intended to show that responsibility does not require an ability to do otherwise writes that, "mental acts such as deciding, choosing, undertaking, forming an intention . . . constitute the basic *loci* of moral responsibility." "Libertarianism and Frankfurt's Attack on the Principle of Alternative Possibilities." *Philosophical Review* 104 (1995): 247–61 at 247.

[52] A recent example of failure to make this important distinction is Roger Clarke's "How to Manipulate an Incompatibilistically Free Agent." *American Philosophical Quarterly* 49 (2012): 139–49.

Moreover, in terms of what is most interesting to the Anselmian, the value of aseity and the character forming nature of choices, the overt action, considered all by itself, contributes little if anything. So, for example, suppose you make an *a se* choice to murder a colleague. You aim carefully, you shoot, and he falls. You believe you have freely murdered him. Unbeknownst to you, he had stumbled just as you shot, and your bullet just grazed his arm. Are you responsible for committing a murder? Not in the eyes of society. No murder occurred. But in terms of the nature of your character, you are a murderer. If we focus on where the real Anselmian action lies, your free choice, you are responsible for committing a murder. In the eyes of an ideal judge—God, for example—the fact that you missed accidentally, while hugely important to your victim, is not very important, or not important at all, in terms of how you are judged. What kind of person you are depends on what you choose to do, not on whether or not you succeed in doing it. (This goes for the good deeds, too. Mother Teresa is quoted as saying that God expects you to try, He doesn't necessarily expect you to succeed.)[53]

This is not to say that the difference between choosing and failing on the one hand and choosing and succeeding on the other is not important. The difference between murder and mere assault (or attempted murder) will deeply affect the story of your future in a number of ways. How we, society, deal with you is certainly not the same on the different scenarios. And how you think about your crime afterwards may be very different on the different scenarios. For example, if you succeed in killing your colleague, you may choose to embrace what you've done, cementing within yourself a murderous character. Or you may be horrified and repent. If you fail to kill your colleague, you might just be angry that you missed. On the other hand, upon seeing him jump up, you might suddenly feel a great wave of relief and decide to repudiate your choice to murder. Your actually going through with the deed shows how firmly you chose it. And your response after the deed will have a profound impact on your character. So the deed itself is surely important. The point is that, in terms of where to locate your moral responsibility and the elements that go into your construction of your character, the actual deed is less important than your preceding act of choice, and quite likely less important than your subsequent choices regarding what attitude to adopt towards what you have done.

[53] The modern reader may find Kant coming to mind, when the point is made that it is the choice, not the overt deed, that provides the locus of responsibility. A comparison between Anselm and Kant lies outside the scope of this work, but I just note here that, while Kant may echo Anselm in some ways, Anselm's eudaemonism sets him far apart from Kant in terms of the proper working of the just will (see Rogers (2008: 66–72)).

This focus on the mental act of choice does raise a significant practical problem for assessing whether or not someone is morally responsible for an act. In the example above, you might claim that you didn't intend to kill your colleague at all. You just wanted to scare him and were aiming well to his side, but he stumbled into the bullet. If you stick to your story, how could we, society, tell any different? Abelard, a younger contemporary of Anselm's, promotes the view that it is the choice that is the bearer of moral significance. He explains that, due to the privacy of mental states, the punishment which society meets out can be justified only as prevention and deterrence. We mere mortals just do not have the information to exact justice.[54] Kant, on the other hand, seems to suppose that we have a clear window into your soul. We must execute the murderer, for example, not for any benefit to society, but because moral logic requires that in willing to kill he chooses to bring death on himself.[55] Kant's supposition seems to be that we have access to the workings of the will of another. Anselm himself does not address the question of public punishment, since his motivation for writing the relevant works on free will has to do with discussing the relationship of God to created agents. Perhaps the best response to this difficulty about public punishment from the Anselmian perspective is to steer a middle course between Abelard and Kant. In judicial systems in Western societies we usually hold that, while deterrence and prevention are important, it is also important that you *deserve* the punishment you get. Punishment is not justified *only as* deterrence or prevention, since, with enough of a backstory, those goals could be achieved by punishing someone who did not deserve it. But we grant that discerning desert is a difficult task with many complexities. We assess the available evidence as best we can and hope to render a fair judgment. This point will be developed at more length in Chapter Five, and other practical problems arising from the focus on the choice, an event which is impossible for the third party to observe, will be discussed as they come up in other contexts.[56] But it is inescapable that, if our concern is your building your character as a free agent, the moral value of overt deeds follows upon the moral value of *a se* choices. Throughout the book, it will be choices that concern us.

Another methodological principle follows from the Anselmian view that being free renders the agent a metaphysically superior sort of thing; for a normal adult human being it would be demeaning if we held that you were not a free agent. In that, *ceteris paribus*, we ought to try to avoid demeaning people, we should not set

[54] *Ethics, or Know Thyself* Bk. 1, 85–8. [55] *Metaphysics of Morals* 6:331.

[56] Even on the assumption of physicalism, the third party must rely on first-person testimony in order to associate an observable brain event with a conscious choice.

the bar too high in describing what it takes to engage in a free choice for which one can bear moral responsibility. That is, we should not require that an agent engage in many and varied extremely sophisticated cognitive activities before we hold that he can make a free choice. For example, Alfred Mele describes the "ideally autonomous" agent as one with a long history of developing many self-reflective properties.[57] Mele's standards are so high that the majority of adult, rational human beings may never be able to achieve the status of "ideally autonomous." But the Anselmian does not want to deny that the majority of rational human agents, in that they fall far short of this ideal autonomy, are incapable of the free will which grounds basic human dignity. The Anselmian happily grants that (for a free agent) the properties Mele lists would conduce to allowing an agent to be more self-governing and that that is a splendid thing. But the Anselmian holds that there are degrees of autonomy and responsibility. In order to choose freely and to have what we can term "basic" responsibility, it is enough that an agent be able to make an *a se* choice. So a methodological principle at work in building the theory of Anselmian libertarianism will be that, *ceteris paribus*, it is better to include rather than exclude more rational adult humans as moral agents. To deny that someone has free will is to demean them, and so in developing the description of an *a se* choice, positing properties that would *exclude* is, at least prima facie, to be avoided. (This issue will prove central to the discussion of self-creation in Chapter Nine, and here the Anselmian will demur a bit from Anselm's own insistence that a fairly sophisticated ability to self-reflect is required in order to make morally responsible choices.)

Another important point is that, while the question of what it takes for an individual choice to be made freely and responsibly will be extremely important, it is the overall character of the human agent that is of primary interest. In the final analysis it is the kind of person you are that counts more than particular choices you have made. The Anselmian takes individual choices to be the building blocks of the edifice of your character. And, in that you are engaged in this project of character construction, how *you* think about your own choices is extremely important. In the free will literature the focus is often on whether or not "we," society, those around you, can or should hold you responsible, or how "we" should react to your behavior. And that is certainly important. But perhaps even more important, from the perspective of the Anselmian project, is how *you* think about yourself and your choices.

Further, in that it is the self-creative aspect of choice that motivates Anselmian libertarianism, your situation during the time *after* your choice becomes very

[57] Alfred Mele, *Autonomous Agents* (Oxford: Oxford University Press, 1995).

significant. The view that your choices produce your character is old and distinguished and is accepted by many contemporary participants in the free will debate, but there is more to say concerning your situation *after* your choice. One area in which this shift of focus can advance the debate is in dealing with the so-called "luck problem," as I will elaborate in Chapter Eight.

Anselmian libertarianism, thus, makes a further contribution to the contemporary free will discussion by opening a new line of inquiry concerning how the agent thinks about, understands, and assesses his choice after the fact. Some contemporary philosophers have mentioned that an aspect of a responsible choice is that the agent afterwards understands the choice to be his own. Robert Kane holds that an aspect of a free choice is that the agent will say, "I'm willing to take responsibility for it one way or the other."[58] But there is considerably more that needs to be said about the attitude of the agent after the act of choice. If choices are indeed character forming, then how the agent assesses the choice is crucial. Suppose you have chosen to do something wrong. Do you admit that you did wrong, repent, and resolve to do better in the future? Do you make excuses? Do you refuse to acknowledge that what you did was wrong, perhaps by refusing to think about it? How you decide to assess a past choice may, in itself, be a free choice. It may be more important in terms of the formation of your character and what kind of person you become than the original choice.[59] This point about self-assessment and self-creation will be one of the issues in Chapter Eight. And, again, it is a consequence of the more fundamental Anselmian approach; the reason Anselm and the Anselmian *want* aseity transcends the importance of our making this or that discreet choice freely. We are interested in our whole lives as metaphysically elevated beings who can contribute to our own creation. This may be one example of how the Christian worldview can color one's intuitions and one's focus. The Christian has concerns about repentance, and forgiveness, and about how things stand with you when you "go to meet your Maker." Thus, on the Anselmian account, consideration of the agent's own subsequent reflections on, and responses to, his choices, and attention to the character that the agent builds by making choices, all play a crucial role in thinking about freedom, responsibility, and related issues. Someone concerned only about how society should treat wrong-doers will likely not find these issues so pressing. And so to a road map of the book.

[58] Kane (1996: 145). Here Kane describes a "self-forming willing" (a choice for which the agent is ultimately responsible) as a "'value experiment' whose justification lies in the future and is not fully explained by the past." This future orientation is rather like the Anselmian approach.

[59] Kane (1996: 125) offers a list of what sort of choices may constitute self-forming willings, but he does not include free choices of self-assessment after a free choice.

Part II: Outline of the Book

Chapter One: Why not Compatibilism?

One key motivation for attempting to spell out a new libertarian theory—for Anselm in the eleventh century and for the contemporary Anselmian—is the belief that compatibilism just cannot give us the free will we want.

A. THE DIVINE CONTROLLER ARGUMENT

A number of contemporary incompatibilists have defended their position by proposing some version of what can be termed a "controller" argument: Isn't it intuitively obvious—as Anselm takes it to be in motivating his theory at the beginning of his *De casu diaboli*—that if someone outside yourself is controlling your act of making a choice, then that choice is not up to you in a way that can ground your responsibility? The contemporary controller argument goes on to propose that natural causes are relevantly similar to a controller, so that same intuition should apply when it is natural causes that have produced your choices. Thus compatibilism fails. Contemporary compatibilists have responded in one or both of two ways. The compatibilist can argue that there is a relevant distinction between a controller and natural causes, so the intuition generated in the controller example does not transfer to natural causation. The compatibilist may also make what I will call the "*tollens*" move; since natural causes do not undermine freedom, so long as the agent has whatever freedom-grounding properties the compatibilist takes to be sufficient, then, if the agent has these properties, a controller does not undermine freedom, either.

Anselm's construction of his libertarian theory is motivated by the controller problem, but of course in his case it is a *divine* controller. If God causes your choice, then it is not from yourself, and you cannot be responsible. Anselm seems to take this for granted. I defend his intuition and argue that introducing a divine controller can provide a stronger version of the overall contemporary argument than has appeared in the literature to date. The distinctions that can be drawn between natural causes and a limited controller do not apply when the controller is God. So, whether or not you suppose that you are living in Anselm's universe, if he is right that a divine control conflicts with free will, then natural necessitating causes equally undermine freedom. And a critique of the compatibilist's *tollens* response serves to strengthen that basic intuition. If we focus—as Anselm does—on your making a wicked choice, the intuition that *you* cannot be blamed if *God* causes your choice is very powerful. And it is hard to see how this intuition is to be dislodged just by adding the compatibilist's list of supposedly freedom-grounding properties when all of these properties are also caused by God. If the concern is aseity, then the divine controller argument is very powerful.

B. A WAGER

I conclude the chapter with the suggestion that, if the divine controller argument does not persuade you, you might want to engage in a little Pascalian-type wagering. A big part of the Anselmian project involves self-construction and this, in turn, depends on how you think about yourself. In the absence of proof either way, the safer bet is to think of yourself as free with libertarian freedom.

Chapter Two: Anselmian Libertarianism: Background and Voluntates

Before I can set out how Anselmian libertarianism describes free choice I need to provide a little background. In that I am concerned to show that Anselm's theory connects with, and yet, in some ways, improves upon, contemporary libertarian theories, it is important to say a word about a few of these theories at the outset—just enough to motivate the comparison with the Anselmian view. First I offer a quick description of Robert Kane's event-causal theory and then of standard agent-causal theories, very briefly noting the major problems contemporary philosophers see with each. Then I set out Anselm's motivation for constructing his theory. Like Robert Kane, Anselm hopes to allow for "ultimate responsibility" on the part of the human agent, but he will not introduce any new, ad hoc, sorts of causes invoked to provide for human freedom. It is this motivation that leads to the parsimony of his approach. Then I set out and defend a couple of preliminary methodological points having to do with how I will be explaining Anselm's theory. First, I will follow Anselm in discussing only choices with moral significance. Second, I will stick with the example Anselm uses, the fall of Satan. Although Satan has not appeared on stage much in the contemporary debates, I will argue that it is helpful to retain the diabolic example for purposes of setting out the Anselmian theory.

Having set out this background material, I turn to Anselm. But I save the actual explanation of his theory of free choice for Chapter Three. In Chapter Two I review Anselm's distinction between three meanings of "will," *voluntas*. Anselm is a careful, analytic thinker, and these three "wills" provide the elements he will work with to construct his theory of free choice. The three need to be on the table first in order, in Chapter Three, to work through Anselm's theory of how *a se* choice is possible.

Chapter Three: Anselmian Libertarianism: A Parsimonious Agent-causation

Chapter Three begins with a discussion of Anselm's understanding of will in lower animals. Anselm has it that there are profound and interesting similarities between animal and human willing. But animals cannot will freely and contribute to their own characters. One crucial difference is that rational agents can be torn between morally significant options. *A se* choice must be preceded by this "torn condition." In that everything about the agent is from God, including the motivating elements that precede a free choice, it is these alternative possibilities that set the stage for the agent to produce something—the choice for this over that—that is truly from himself.

But Anselm is insistent that the *a se* choice does not require any special causal powers unique to human choosing. Anselm argues that, while it is absolutely up to the agent to cause the choice, the agent causes the choice by "per-willing" one God-given desire over another. The term is coined in this context by Anselm himself and means roughly "to will through to the point of intention." This per-willing occurs in a situation in which the agent is torn between two genuinely open alternative possibilities. He desires two incompatible, morally significant "objects," and the per-willing of one entails overriding the desire for the other. So the agent causes the non-determined choice, but simply by per-willing, and that means without the introduction of any new sorts of causes. The adherent of a standard contemporary agent-causal theory might argue that Anselmian libertarianism is *too* parsimonious in this regard. He might complain that it does not ascribe enough causal power to the agent. I conclude the chapter by offering a brief response to this criticism.

Chapter Four: Three Entailments

In this chapter I spell out some of the entailments of Anselmian libertarianism.

A. THE ONTOLOGICAL STATUS OF CHOICE

A consequence of the Anselmian theory is that the act of choice is not any sort of "thing" with ontological status. This is exactly the conclusion that Anselm himself was aiming for, since it conforms to his classical theism, but prima facie it is puzzling. I attempt to explain and defend the thought that a choice is not a "thing." Rather it is a "thin" event completely dependent for its "being" upon other things and events, and so having no ontological status of its own. I note some practical consequences of this position.

B. THE GROUNDING PRINCIPLE

Anselm's insistence on the absolute aseity of a free choice entails what I label "the grounding principle"; what grounds the truth of a proposition about a free choice, and what constitutes the source of knowledge concerning a choice, is only the agent actually making the choice. Some (self-styled) libertarians, such as contemporary Molinists, flatly disagree and hold that the truth and the knowledge can and do exist independently of the choice. On Anselmian libertarianism this view is just incoherent. Some libertarians, in discussing Frankfurt-style counterexamples, come close to proposing the grounding principle or seem to assume it. But, to my knowledge, the principle has not been clearly stated in the free will literature. (I believe there are some metaphysicians who hold to the principle or something very like.) Anselm, because he proposes his libertarianism in a universe in which he supposes there to be an ideal knower, who knows even future free choices, presents the principle with a clarity which is helpful in solving problems raised in the contemporary discussion.

C. CHARACTER CREATION

Anselm proposes libertarianism in defense of the elevated metaphysical status of the created agent as a self-creator. He follows in Aristotle's footsteps, arguing that you create your character through the choices you make, and that, if your self-formed character determines your subsequent choices, you are nevertheless responsible for those choices. (This move is common in the contemporary literature, but it is helpful to get it on the table in its Anselmian version.)

(Chapters Two, Three, and Four explain the theory. Chapters Five through Nine defend it against several possible criticisms drawn from the contemporary free will debate.)

Chapter Five: Defending Anselmian Internalism

The Anselmian holds that the criteria requisite for a free choice are factors which concern the choice itself and its immediate past. This sort of position raises problems. For example, one of the main criticisms which Alfred Mele levels against Robert Kane's libertarianism is that it focuses on the structure of the immediate choice—Mele labels this internalism—rather than on the history of the choice. The Anselmian, too, focuses on structure, rather than history, when the choice in question is an *a se* choice, as opposed to a character-determined choice. Mele notes that, on Kane's account, the agent is said to be free, although he may not have had any control over which competing motivations he is struggling to realize. But then, says Mele, even if there is some indeterminism in the choice, it does nothing to enhance the agent's autonomy. Mele defends his criticism by proposing examples in which the alternatives which the agent confronts in a Kane-type choice are suddenly supplied by some covert controller and are not produced by, or consistent with, the agent's history. Mele takes it that our intuition will be that the Kane agent may well choose in a non-determined way, but that he does not have any freedom worth wanting.

Prima facie, Anselm's use of the example of Satan, where it is clearly stated that God, and not Satan, is responsible for everything about Satan's past, might suggest that Mele's criticism is even more telling against the Anselmian theory than against Kane's. In fact, Kane's theory is more susceptible to this criticism than is Anselm's because of the difference in where they locate the indeterminism in the choice. Nonetheless, the criticism still has some purchase, in that Anselm is quite clear that we are responsible for our choices, even if the motivating factors were completely outside our control.

I agree that on the Anselmian theory an agent may choose freely, and yet, at least with choices early in life, have very little real autonomy. However, I argue, using examples analogous to Mele's, that the alternative possibilities in question do indeed allow the agent to choose with an aseity worth wanting. At the very least the Anselmian agent is more autonomous than Mele's determined, "autonomous" agent. This defense against Mele provides a good example of the point that background assumptions may be playing a role as philosophers construct examples and experience intuitions. Anselm and the Anselmian are working from a background in Christian culture in which stories of conversion abound. Conversion involves a sudden break with one's past, which yet leaves the agent's free will standing. If narratives of conversion play a constitutive role in one's understanding of the human condition, then one is less likely to share Mele's intuitions that drive his criticism, or so it seems to me.

Internalism does generate some problems. First, it is at least theoretically possible that an adult agent's character is determined by free choices he made in his youth when he had little if any control over what factors would motivate him. Second, granting that the agent may have little control over what has brought him to the point of being in the torn condition, it is reasonable to take this into account when we must assess the degree of an agent's responsibility. But this is extremely difficult from the third-person perspective. I attempt to mitigate these difficulties.

Chapter Six: Anselmian Alternatives and Frankfurt-style Counterexamples

Anselm's libertarianism holds that the created agent must confront genuinely open options in order to be able to choose with aseity. Decades ago Harry Frankfurt popularized a criticism of the libertarian's requirement of alternative possibilities. (Frankfurt attributes this criticism to Mill, but, interestingly, a version of it—regarding only overt actions, not choices—is found in Anselm's work.) It goes roughly like this: An agent is debating whether to do A or B. Some covert controller is ready to step in, if he sees that the agent will choose to do

B over A, and will bring it about that the agent chooses A. As it is, the agent chooses to do A over B on his own. The controller never steps in. Isn't the agent free? Since Frankfurt first posed this criticism, the debate has gone back and forth with many developments and nuances, including (as prefigured in Anselm) a careful restriction of the relevant alternatives to the agent's choices, not the agent's overt actions. Much of the debate has revolved around how, and when, and if, the controller could foreknow what the agent will choose. Recently Alfred Mele and David Robb have tried to strengthen the criticism by reposing it without the controller. Their "blockage" example posits, instead, a rather puzzling and convoluted "process" that makes the agent choose A if the agent is going to choose B over A.

Anselm's commitment to the grounding principle, and to an ideal and omnipotent knower, brings clarity to the issue. Even God cannot make it the case that an agent who is going to choose B, chooses A. God knows future free choices. But, as the grounding principle entails, He knows that the agent chooses B at a time (t1), only because the agent chooses B at t1. It is logically impossible that an agent choose B at t1 and fail to choose B at t1. Even God cannot bring about the logically impossible. The original set up of the Frankfurt-style counterexample is just incoherent on Anselmian libertarianism, and no amount of tinkering, including replacing the proposed controller with a "process," can rework it so as to connect with Anselm's requirement for alternative possibilities. If God can't make an agent who is going to choose B, choose A, no mere "process" can do it!

Another recent attempt to reframe Frankfurt's criticism, proposed by David Hunt, involves "buffered" cases, where the controller will step in at an earlier stage of the decision-making process than the point where the agent is debating between choosing A and B. Roughly, if the agent starts to consider option B, as well as the controller's desired A, then the controller will bring it about that the agent does not consider B. The agent, on his own, never considers B and chooses A, and the controller does not step in. Isn't the agent free? On Anselm's account, obviously not. In that Hunt's agent is never in the torn condition, the situation which allows for an *a se* choice does not arise.

There looks to be one conceivable scenario, "Rewind," in which a controller can see to it that an Anselmian free agent chooses what the controller would have him choose, but it involves rewinding the universe and so is not a Frankfurt-style counterexample. It is probably metaphysically impossible in any case. And—at least possibly—Rewind renders the "controlled" choice of the agent ENC necessary, and hence not *a se*.

Chapter Seven: The Luck Problem: Part I. Probabilities and Possible Worlds

Perhaps the biggest problem which contemporary philosophers see with libertarianism is the luck problem, and so I devote Chapters Seven and Eight to it. Seven addresses a series of preliminary issues, and Eight offers an Anselmian response to the luck problem. The problem is this: If an agent's choice is undetermined then there is nothing about an agent's past, including his character, that explains why he chose B over A. It is absolutely the case that he might just as well have chosen A over B. But then isn't his choice just a matter of luck? And if that is the case, how can the agent be held responsible for this "lucky" event? Anselm himself recognizes the cognitive discomfort of having to insist that there is no preceding cause or explanation for why the agent chose B over A. But Anselm cannot allow the alternative that God causes all choices, since that would contradict human freedom and responsibility. And, if the divine controller argument of Chapter One is correct, determinism by natural causes entails the same unwanted consequence.

I allow the cognitive discomfort, but in this chapter I begin an effort to weaken the luck problem as it appears in the contemporary literature, and especially in the work of Alfred Mele. Chapter Seven, then, makes little appeal to the work of Anselm himself, but the Anselmian must address the problem, in that, if it succeeds against libertarianism in general, it succeeds against Anselm's version. In Chapter Eight, though, I will appeal to a particular move Anselm makes which, I believe, constitutes a significant advance in answering the luck problem.

The first part of Chapter Seven is devoted to a discussion of the sort of rhetoric used by libertarians and anti-libertarians alike that helps to fuel the intuition that a libertarian free choice is merely lucky. Some philosophers pose the luck problem by describing the libertarian choice in ways that suggest that it is like throwing a die. Mele, for example, describes it as a mental spin of the roulette wheel. Van Inwagen suggests that, were God to rewind the universe numerous times, the choice would fall out for A about half the time and the same for B, showing that the choice for B has a preceding probability of 0.5. And doesn't that sound like mere chance? He notes that the number is arbitrarily chosen to make the point, but still, if choice is just a matter of a probabilistic chance, then it seems to be a mere matter of luck. Robert Kane, and other libertarians, invited this criticism when they described the choice as "probabilistically caused." But the Anselmian not only denies that the choice is probabilistically caused, the Anselmian insists that the very concept of anything remotely resembling the sort of "numerical probability" that science employs—whether probabilities are understood to involve propensities or simply relative frequency—is inapplicable to a libertarian choice. Insofar as the "probability" talk helped drive the anti-libertarian intuition, showing that there are no numerical or scientific probabilities (beyond "more than 0 and less than 1") involved in libertarian choice should undermine the intuition.

Recently Alfred Mele has suggested that couching the luck problem in the language of possible worlds enhances its intuitive appeal: That the agent is in this world in which he chooses B is just a matter of luck, since he could equally well have been in the other possible world in which he chooses A. I consider what the possible worlds talk really means and conclude that Mele's casting the luck problem in the possible worlds locution does not add anything of significance to the original statement of the problem.

There are two well-known and popular analyses of what a possible world actually is: (1) Possible worlds are consistent representations of how a world might be. (2) Even non-actual possible worlds exist as much as the actual world (David Lewis). The former analysis just seems to say that holding that the agent "could equally well have been in the other possible world" means only that his doing otherwise, given the same history, can be consistently represented. This is indeed what libertarians say, but it is difficult to see that this adds anything that should strengthen the anti-libertarian intuition. The latter analysis apparently entails that agents "choices" are ENC necessitated, and so agents are not free. Or, to put it conversely, on the assumption of Anselmian libertarianism, there is not a (really existent) possible world in which the agent, who freely does B in the actual world, freely does A. Mele's possible worlds version of the luck problem does not seem to add anything to the original version which should enhance anti-libertarian intuitions. Thus I turn to the classic statement of the luck problem in Chapter Eight.

Chapter Eight: The Luck Problem: Part II. The Locus of Responsibility

Some argue that we praise and blame the person for his character, and if the choice is not a consequence of preceding facts about the person and his character, but simply a fleeting event, then the agent is not responsible for it. Hobart, in the 1930s, emphasized the point about holding the person, with his preceding character, responsible. Hume prefigures that point, but notes, as well, the ephemeral nature of the choice, which does not, he says, leave any trace behind it. Anselm holds that Hume is wrong and that Hobart has the relationship of character to responsibility and choice backwards.

The whole point of being free, in Anselm's view, is that we should be able, by our choices, to help in the construction of our own characters. The event of a choice may be fleeting, but it can have an enormous impact on our character. What is blameworthy or praiseworthy is not the character which precedes the choice, but the choice itself and the character it creates. The thought that we develop our characters through our choices is at least as old as Aristotle, but the present free will debate focuses almost exclusively on the choice and its history. The Anselmian proposal is valuable, not just as a new way of addressing the luck problem, but also as an encouragement to consider the effects of a choice on the future of the agent.

One interesting issue related to the future of the agent is the question of forgiveness. On many views about the justification for punishment and the relationship of choices and deeds to character it is difficult to justify forgiveness. The Anselmian thesis that what we are concerned with is the present character produced by the past choice allows for an explanatory framework for standard intuitions about forgiveness—standard at least for those with a background in Christian culture.

Chapter Nine: The Tracing Problem

As discussed throughout this work, including especially in Chapter Eight, the thought that we must have libertarian freedom in order to be self-creators is central to Anselm's theory. One important entailment of the ongoing project of self-creation is that we can be responsible for our character-determined choices if we created that character through earlier, *a se,* choices. Responsibility can be "traced" back to those choices. Anselm takes it for granted with nothing in the way of argument or defense, and many contemporary libertarians, Kane for example, also make the tracing move. But serious problems can be raised against this tracing thesis. In that self-creation is the purpose of freedom in Anselm's view, these problems must be addressed by the contemporary Anselmian who hopes to develop and defend the theory.

It is commonly assumed that, in order to be responsible for a choice, you must have some understanding of what it is you are choosing and doing and of what the consequences of your choice will be. Call this the "epistemic requirement." But in making choices, especially those early choices which may be so important to the formation of character, do we recognize that we are *indeed forming our characters*? If not, how can we be responsible for our characters? (To my knowledge, the issue of character-forming per se has not been much discussed in the contemporary free will literature. A different, but related, question concerning our ignorance of the specific, future situations that we may face has come up. I discuss this latter question briefly, in that it is easier to address and has been, I believe, successfully answered already.) There are two, mutually consistent, avenues which the Anselmian can pursue in addressing this problem.

A. ARISTOTLE'S ANSWER

One answer, which some contemporary philosophers defend, and which is usually attributed to Aristotle, is simply that any agent capable of making a free choice does, or at least *ought*, to grasp that his choices affect his character. This is not an implausible claim. Prima facie one might suppose that sophisticated self-reflection is required in order to understand that one creates one's character through one's choices and actions, but common experience teaches otherwise. Even parents who have never read, or even heard of, Aristotle know that the way to instill the habits of virtue in small children is to encourage (or force) them to behave well under the guidance of those who already have the skill. And—to note, again the importance of background worldview to issues in the free will debate—anyone brought up in the Christian tradition has, presumably, learned self-reflection from a very early age. Repeatedly petitioning God the Father to "Lead us not into temptation" entails consideration of one's own motivational states. So the Aristotelian supposition that most agents do, or should, understand character-creation is defensible. Nonetheless, there may be those—such as young children or the congenitally un-self-reflective—who do not see, even on some very basic level, that their choices inform their future characters. We might just exclude such folk from the circle of morally responsible agents, but in that it is a great thing to be a morally responsible sort of being, we should not deny that status to anyone unless we must.

B. QUALIFYING THE EPISTEMIC REQUIREMENT

Explaining the "epistemic requirement" I wrote, "It is commonly assumed that, in order to be responsible for a choice, you must have some understanding of what it is you are choosing and doing and of what the consequences of your choice will be." But this statement is open to qualification. First, it seems obvious that you do not need to be aware that you are engaged in the invisible action which philosophers refer to as "making a choice." That is, in order to be responsible for what you choose, you do not need to have "stepped back" cognitively and taken note of the fact that you have made or are making a choice. Ignorance that you are making a choice is not exculpatory. The ignorance which is exculpatory is ignorance concerning the *content* of the choice, what the choice is "about." And this same distinction can be invoked regarding character-formation. Exculpatory ignorance involves innocently not knowing something concerning the content of the character-forming choice. Not recognizing that you are in fact making a character-forming choice is a different species of ignorance and does not undermine your responsibility for the choice, or for the subsequent character.

In conclusion, the Anselmian theory is defensible against the standard criticisms leveled against libertarianism. It provides for the agent's control, just where it is needed. We human agents can choose with the aseity required for responsibility, and so we can have the elevated metaphysical status that the Western tradition ascribed to us. And the Anselmian achieves this with a parsimonious theory that does not add any ad hoc new sorts of things to the universe.

1

Why not Compatibilism?

The Divine Controller Argument and a Wager

> But there could be no reason why God would justly reward the good and the evil according to their individual deserts, if no one had done either good or evil through free will.
>
> *De Concordia* 3.1

Introduction

The intuition that motivates Anselm's entire libertarian project is that created agents cannot be free and responsible, cannot be genuine *imagines dei*, if their choices are ultimately causally traceable, directly or indirectly, to God. (For the purposes of this chapter the question of external non-causal necessity can be bracketed.) Anselm himself devotes a great deal of time and thought to explaining how it is possible, in the universe of classical theism in which God is the cause and constant sustainer of all that exists, that the choices of the created agents could be up to them, and not to God. Anselm's theory is the subject of Chapters Two, Three, and Four of the present work. Anselm does not attempt to defend the basic intuition. One gets the distinct impression that he is writing for people who share it.[1] In the present chapter I hope to go some way to defending the intuition, by stating it as clearly as I can, spelling it out with examples, and by confronting contemporary compatibilists who dispute it. I will do this within the context of a particular move in the contemporary free will debate, the controller argument. This argument was introduced to defend incompatibilism, and its

[1] His *De libero arbitrio* (DLA) and his *De casu diaboli* (DCD) are both dialogues in which the student and the teacher discuss problems of free will. The "student" may well be based on Anselm's own students. In any case, the student does not dispute the thought that if choices are traceable to God, such that the agent must make the one choice which God determines him to make, then they are not free.

initial premise is similar to Anselm's basic intuition about God and free choice. One move made by those who attack the controller argument in defense of compatibilism requires them to conclude that you may indeed be free and responsible, even if your choices are controlled. I attempt to portray the weakness in this move, and, in so doing, support Anselm's basic intuition. In that it is a basic intuition, I do not see that there is much more to be done. (Now that universal determinism is no longer held to be the "scientific" tack to take, those philosophers who reject the basic libertarian intuition often do so after having come to hold that there are fatal flaws with libertarianism. Thus my attempts to defend Anselmian libertarianism against some standard criticisms in Chapters Five through Eight can be seen as a sort of defense of the libertarian intuition.)

In the present chapter I will use the intuition about a *divine* controller to strengthen the contemporary controller argument for incompatibilism.[2] The standard controller argument usually targets universal or natural determinism. Though Anselm himself is not concerned about natural determinism, his thesis is clearly that if an agent chooses by causal necessity, then the choice is not *a se* and hence not robustly free. From the perspective of the contemporary Anselmian all of these species of causal necessitation, divine and natural, need to be addressed, and so I will devote some time to that aspect of the controller argument that deals with universal determinism. If my divine controller argument works, it should persuade any compatibilist—theist or not—to abandon his position in favor of incompatibilism.

The contemporary incompatibilist begins his controller argument with the assumption that it is intuitively obvious that if a controller—typically a mad neurosurgeon or a character of that sort—makes you choose what you choose, you cannot be held responsible. But, continues the controller argument, natural causes are relevantly similar to the hypothesized controller, so if it is natural causes that produce your choices, you are not responsible. There are two standard responses the contemporary compatibilist makes. First, natural causes are *not* sufficiently similar for the argument to go through. Were that response effective, then freedom and responsibility would be safe in a determined universe, so long as there are no controllers controlling. Second,—I will call this the *tollens* move—*if* natural causes are relevantly similar to the controller, then perhaps you *are* responsible, even when the controller causes your choices. This is the response more relevant to Anselm's own thinking, since what he is actually concerned to avoid is the theist compatibilist position that God *does* control the choices

[2] I argue this point at more length in "The Divine Controller Argument for Incompatibilism." *Faith and Philosophy* 29 (2012b): 275–94, from which this section is drawn.

of created agents and yet these agents are free and God holds them responsible for what they choose. If the *tollens* response succeeds in undermining the controller argument, then the driving intuition behind Anselm's theory is seriously weakened.

In the present Chapter I will canvas a number of definite advantages to running the controller argument in an Anselmian spirit with God in place of any old controller. Most importantly, with God in the controlling role, all of the proposed distinctions which are supposed to show that the controller is relevantly unlike natural causes can be nullified. The judgment regarding human freedom and responsibility ought to be the same, whether it is nature or nature's God in the driver's seat. I then address the *tollens* response. I note that the compatibilist has a difficult task in that the intuition that you are *not* responsible for your divinely caused choices is almost certain to be stronger than any intuition that you *are* responsible even though you are naturally causally determined to choose as you choose. I argue that no amount of additional (but divinely caused) elements added to the history or structure of a choice can supply the aseity that is the foundation of responsibility. The upshot is that the debate over the controller argument ultimately serves to strengthen the basic Anselmian intuition.

For those unconvinced by the divine controller argument I add a brief wager. It appeals to the old argument—not looming large in Anselm's work, but likely familiar in his day—that belief in determinism is harmful in that it is likely to encourage moral laxness. My hope is that, by casting it as a wager, I have expressed this old argument in a lively and persuasive way.

The Controller Argument and Advantages of Including God

In the contemporary literature a number of "manipulator" or "controller" arguments have been advanced against compatibilism.[3] A typical controller argument goes roughly like this: Hypothesize a controller—a mad neuroscientist or a megalomaniac behavioral engineer—who causes you to choose to do something (X) in such a way that your choice is necessitated. (Note again that "cause" should be understood in a broad, and stronger than counterfactual, sense. To cause something is to exert some force or power to produce an effect. The controller

[3] Recent examples include Robert Kane, who proposes the argument as part of his defense of libertarianism (1996: 64–71) and Derk Pereboom, who uses it to bolster a medium-hard determinism ("Determinism al Dente." *Noûs* 29 (1995): 21–45). These arguments have also been called "manipulator" arguments, but I will argue that "manipulation" is not the right term to describe *divine* control.

brings it about. Were the causation in question a Humean constant conjunction, then the very thought that the controller *controls* would be problematic, and one could not really mount a controller argument.)[4] Your choice to X comes about in such a way that you could not possibly fail to choose to X. And suppose that the controller controls without coercion, and you do not even know of his activity. To adopt Robert Kane's terminology, he is a "covert, nonconstraining controller" (CNC).[5] You might "feel" free, but nonetheless, the argument goes, it is intuitively obvious that you cannot be held responsible for the choice to X. Why not? Because someone else made you make the choice. But, continues the incompatibilist, analysis of the controller hypothesis shows that what precludes your having moral responsibility is not so much that there is someone else involved in your choice, but that their involvement causally necessitates your choice. If your choice is causally necessitated by something other than yourself, then you cannot be responsible for it.[6] But if universal determinism is true, everything is causally necessitated. (As noted in the introductory chapter, it has been common to focus on *universal* determinism. In that the contemporary controller argument has been cast in terms which involve universal determinism I will put my discussion in those terms as well. But the free will debate continues roughly the same if the sort of determinism in question includes only natural determination of human choices.) Your choices are the inevitable product of something not yourself. So, in a deterministic universe, no human choices are free in a sense which can ground moral responsibility.

In response, the compatibilist who agrees with the initial intuition concerning the controlled agent's lack of responsibility may reject the claim that the causally necessitating factors at work in a deterministic universe are the same as, or relevantly similar to, those in the controller scenario, and so he can argue that the conclusion need not follow. In the current literature this is sometimes referred to as the "soft line" response. Alternatively, if the compatibilist grants that the controller as cause and the deterministic universe as cause are the same or relevantly similar he can conclude—the "hard line" response—that the original intuition was misleading. Just as we can be free and responsible in a determinist universe, we can be free and responsible when our choices are caused by a controller. If this leaves us with an intuitive draw, it can be argued that the

[4] Rogers (2012b): 291–3. [5] Kane (1996): 65.

[6] The qualifier, "by something other than yourself" is required to take account of situations in which you have freely and responsibly constructed a situation such that your choice is causally necessitated by that situation. For example, Anselm holds that if you produce your character by *a se* choices, and then your present choice is determined by that character, you made the choice freely and responsibly.

incompatibilist loses, since it was he who proposed the controller argument to discredit compatibilism.

Locating this argument within the Anselmian universe immediately suggests putting God in the place of the limited controller who is the standard character in the contemporary literature. This move has been mentioned recently, but it has not been developed.[7] Positing a divine controller improves the argument in a number of ways, one of which deserves mention at the outset. Like so much of the free will discussion, the dispute here between compatibilists and incompatibilists is a battle of intuitions. I think it likely that the controller argument provides an example of the phenomenon I mentioned in the previous chapter; our intuitions regarding free will questions may be informed by our positions on more general or more fundamental issues. That was why I preferred not to make a pretense of bracketing Anselm's theism in the process of developing a theory of free will based on his work.

In standard controller cases intuitions may be influenced by the fact that the limited controller, such as the mad neuroscientist or the megalomaniac behavioral engineer, are the inventions of philosophers to serve a fleeting purpose. Intuitions concerning the divine controller argument may play out differently in that Anselm's God has a history and a following such that He must be taken seriously. The limited controllers are "thin," bizarre, and evanescent, which makes them suspect as "intuition pumps." By "thin" I mean that *all* there is to them is their role in the controller argument. They are bizarre figures, so it can be argued that their weirdness is doing much of the heavy lifting in eliciting the looked-for intuition. And they are evanescent. We do not *really* have to worry about them since they exist only as fanciful hypotheses, safely confined to the pages of philosophical literature. As we do our philosophy, we may commit to consequences involving these characters because we believe that it is safe to do so. The consequences, like the characters, will not "slop over" into "real life." However, if we, like Anselm—or even just most of our forebears and many of our neighbors—believed the controller to exist in reality, then we might be more cautious about the conclusions we accept regarding the controller scenario.

Anselm's God, on the other hand, is not "thin." The Anselmian concept of God is systematic and complex, so intuitions in His regard will not be based solely on a narrow role in the controller argument. And, while God may be bizarre in the

[7] For example, Derk Pereboom quotes a devoted Calvinist in drawing out and making vivid the entailments of insisting that we may be free and blameworthy though controlled ("A Hard-line Reply to the Multiple-Case Manipulation Argument." *Philosophy and Phenomenological Research* 77 (2008): 160–70, see especially 165–7). But he does not make the divine controller case the focus of his argument.

sense that He is a very unusual sort of person, He is nonetheless a common and well-known figure in Western thought. And the idea of God has an importance that the idea of the mad neuroscientist and the megalomaniac behavioral engineer do not. Believers or not, our intuitions regarding the God hypothesis are correspondingly likely to be more serious. Of course, it may be that, to some extent, and for some philosophers, intuitions will just differ between theists and non-theists. Non-theists need not worry that God might *really* be controlling human choices, while theists could have a genuine concern. And that big difference in how they view the world might translate into firm and differing intuitions regarding the divine controller hypothesis.

The *Divine* Controller Argument

1. If God determines your choice, then you are not morally responsible for it.
2. Natural determination of your choice due to natural causes in a deterministic universe is relevantly similar to divine causal necessitation.[8]
   Therefore
3. If natural causes in a deterministic universe determine your choice, you are not morally responsible for it.

But why does God's making you choose conflict with your responsibility? Anselm responds that the reason is that, in such a case, the decisive causal impetus necessitating your choice does not come from you, but from God. It is not under your control and precedes your choice logically and perhaps temporally. However we phrase it, Anselm's demand for aseity has at least a prima facie plausibility. Suppose Premise 1 does not have any intuitive purchase on you. Suppose you are comfortable with saying, for example, that God may cause you to choose to commit a murder and then hold you morally responsible for it. He may justifiably blame you for it and perhaps even punish you. If you do not find Premise 1 intuitively plausible, then the divine controller argument will not impress you, and you can rest easy being a compatibilist. The Anselmian finds it demeaning to deny that human choices are *a se*, and just unbearable to suggest that God should cause all your choices, including the wicked ones, and yet justly hold you responsible.[9] So from an Anselmian perspective we can develop the basic intuition driving the divine controller argument a little more:

[8] Which is not to say that divine causation is much like natural causation. All I need for the argument is that, as I will argue, they are similar in that both can causally necessitate an effect.

[9] In DLA 8 he argues that it is logically impossible that God could cause the created agent to sin.

1. If God determines your choice you are not morally responsible for it *because* the choice is not *a se*. (The causal impetus necessitating your choice is not under your control and precedes your choice logically and perhaps temporally.)
2. Natural determination of your choice due to natural causes in a deterministic universe is relevantly similar to divine determination.
   Therefore
3. If natural causes in a deterministic universe causally necessitate your choice, you are not morally responsible for it.

Compatibilism is false. Q.E.D.

Are the Controller and the Natural Universe relevantly Similar?

The compatibilist will respond that this is far too fast. For one thing, there may be a relevant difference between the divine controller and natural determinism such that Premise 2 is false and so—as long as we do not live in a theist universe with a divine controller controlling all choices—the conclusion that there is no responsibility in a determinist universe can be blocked.[10] This is one of the critiques Alfred Mele levels against Derk Pereboom's controller argument, the four-case manipulation argument. [11] Pereboom presents four cases where agents' choices are determined.[12] He begins with a limited controller case where the controllers directly manipulate an agent's mental processes to produce a given choice. He holds that it is intuitively obvious that the controlled agent is not responsible in this case. Then he moves progressively through three more cases. The second involves indirect control through programming. In the third, control is exercised through rigorous training. The fourth results in the same consequences for necessitated choice on the part of the agent as in the third, but the causes are natural, not produced by a controller. Pereboom argues that, just as it is intuitively obvious in the first case, the three succeeding cases are sufficiently similar that they should elicit the intuition that the agent is not responsible, and this includes case four, where natural causes produce the agent's choice in the determinist universe.

[10] Bernard Berofsky, "Global Control and Freedom." *Philosophical Studies* 131 (2006): 419–45; David Blumenfeld, "Freedom and Mind Control." *American Philosophical Quarterly* 25 (1988): 215–27; Bruce Waller, "Free Will Gone Out of Control." *Behaviorism* 16 (1988): 149–57; Gary Watson, "Free Action and Free Will." *Mind* 96 (1987): 145–72.

[11] Alfred Mele, *Free Will and Luck* (Oxford: Oxford University Press, 2006): 138–44.

[12] Pereboom (1995).

Mele does not try to pinpoint exactly what the relevant difference between the manipulator and natural determinism is. I will argue that the possibly relevant differences between a limited controller and the determinist universe disappear when the controller at issue is God. So what might the relevant differences be? In the limited controller case in which the mad scientist causes your choice, especially if it is a choice to do something wrong, we might hold that there is something morally wicked, or at least suspect, in the controller's behavior. And couldn't the presence of a blameworthy agent—other than you—in the history of your choice, drive our intuition that you should not be held responsible? With the *limited* controller this point has traction. We have, after all, hypothesized a *mad* scientist or a *megalomaniac* behavioral engineer. History demonstrates that when mere mortals put themselves *in loco divinitatis* and attempt to control their fellows, things go badly for the would-be controlled. Perhaps these thoughts form part of the background of our intuition that the controlled agent is not free. And since natural causes in a determinist universe cannot be accused of moral turpitude, we have reason to believe that the causation exercised by the controller is relevantly different from the natural causes at work in a determinist universe. But with the divine controller what drives the intuition cannot be moral qualms about the behavior of the controller. Anselm's God is necessarily good, so our unwillingness to ascribe moral responsibility to you when God causes you to choose to murder (for example) cannot arise from our holding that God has behaved badly and so must bear all, most, or at least some, of the responsibility.[13]

Nonetheless we may hold that, even if we have to allow the controller's goodness *ex hypothesi*, you are still being used by another agent, and it is resentment at the thought of being used that drives our intuition that you are not responsible in the divine controller case. The deterministic universe has no purposes, so it is not *using* us, and the situation is relevantly different. But we can alter the divine controller case to ensure that you are not being used in ways you could properly resent. We could say that, before God causes your choice to murder, He shows you his plan, and you, seeing the benefits that will ultimately be produced by your crime and punishment, agree to allow him to cause the choice. (If your agreement is not freely given we may be opening the way for an infinite regress, so perhaps we could add here that God has bestowed on you an ability to make a libertarian free choice for just long enough for you to choose to

[13] Some people seem to be terrified by the thought that there is a God, and so perhaps in their case, the relevant difference that they see between divine causation and natural causation is driven by what we might refer to as "theophobia." This terror does not seem to be a rational response to the suggestion of theism, and so I do not take the theophobic intuition of a relevant difference to constitute a serious objection to the divine controller argument.

agree that God should cause you to make the choice to murder.) Then He erases your memory and causes the choice—but you agreed to it, so you are not really being used.[14] Or suppose—contrary to the fact, alas—that God had made our world one in which good deeds are rewarded with earthly benefits. He makes you do something extremely good and reap the significant rewards. In this case we cannot point to a difference between the divine controller and natural causation based on the premise that you resent being used by the divine controller. You're happy as a clam, and it would be odd for you to resent it!

Someone might argue that, nonetheless, in both of these examples you are still an instrument in the divine plan. You are being used for a purpose, and that indicates an ineradicable difference between the divine controller and the deterministic universe. So change the example just a little more. God causes you to choose to murder (or to do the extremely good deed) for no reason at all. There is no plan or purpose. This proposal constitutes a departure from Anselm's God, but the point here is to show that the causal necessitation which undermines freedom is still in play even if we subtract from the controller that He has a plan in mind when He causes your choice. Those unfamiliar with the history of Western theism might suppose that a God who wills something, without there being a good reason for it, is a fanciful invention, kin to the mad neuroscientist and the megalomaniacal behavioral engineer. But no. There is a major strand in the philosophy of religion, going back at least to Alghazali in the eleventh century and ably represented by William of Ockham, that insists upon the primacy of the divine will, even above the divine intellect. God, according to this school of thought, is unqualifiedly free to choose anything logically possible. The Good is whatever God should choose. Thus his will is necessarily "good" and is not constrained by anything. He may cause a choice in a created agent without this being "in order to" achieve anything. He just causes it. So the created agent is not being *used* as a means to an end any more than if his choice were caused for no purpose by a deterministic universe.[15]

[14] I grant that this is an odd case. Could we argue that, since you agreed to have it caused in you, you are responsible for the choice to murder? But the situation we have envisioned is one in which the choice to murder, if it is a responsible choice, is *blameworthy*. Are you properly blamed for the murder and simultaneously properly praised for agreeing to have the choice to murder caused in you and so to suffer as an instrument of the divine plan? For a similar, but even stranger, suggestion see Alvin Plantinga, "Supralapsarianism, or 'O Felix Culpa'." In *Christian Faith and the Problem of Evil*, edited by Peter van Inwagen (Grand Rapids, MI: William B. Eerdmans Publishing Company, 2004): 1–25.

[15] It could be claimed that the divinely controlled agent is different from the agent in the deterministic universe because he has been uniquely "singled out." His choices are not caused in the "normal" way. That is easily answered by hypothesizing that every non-divine agent is a divinely controlled agent.

So we cannot ascribe morally doubtful qualities to the divine controller, and we can construct scenarios in which the controlled agent is not being used. Still, isn't there, in the controller scenario, whether we are talking about a limited or a divine controller, an inevitable element of *intervention,* of *manipulation,* by the controller, which is just not there in the deterministic universe? With the limited controller it would seem that some sort of intervention or manipulation would have to be part of the scenario, but not necessarily with the divine controller. Intervention and manipulation, I take it, imply that the controlled agent exists independently of the controller such that the controller must "step in" and tinker with the agent. It could be suggested that someone who intervenes or manipulates introduces changes which turn the agent from the path he likely would have followed. But suppose we take our divine controller to be similar to Anselm's God. Though Anselm denies that God is the cause of human free choices, He subscribes to the God of St. Augustine and St. Thomas Aquinas, the God of classical theism.[16] If we posit that it is this God who is our controller, then the control in question does not entail any intervening per se. Classical theism holds that God's creation consists in sustaining everything in being from moment to moment. Absolutely all that has any ontological status is immediately caused by God simultaneously with its existence.[17] Thus nothing which is not God—no object, no positive property, and thus, given our present controller hypothesis, no action—exists independently of God's directly causing it.[18] (Clearly, in attempting to make room for created aseity, Anselm has his work cut out for him! The key will lie in his holding that a choice per se has no ontological status, a proposal that will be discussed in Chapter 4.)

Classical theism does not deny the sorts of causes that science describes. The causal connections we observe in nature are real and play the explanatory role which science ascribes to them. To pick a standard medieval example, we can observe fire burning cotton. If we then ask, "What caused the cotton to burn?" the correct answer is "the fire." But all of the objects with their properties, powers, and behavior—the whole system of cause and effect—is kept in being immediately by God. In medieval parlance, God is the "primary" cause and the natural

[16] Rogers (2008): 16–29.

[17] Anselm would take it that a metaphysics in which a cause must, by definition, precede its effect temporally is deeply misguided.

[18] Even the laws of logic and mathematics do not exist independently of God, but rather are reflections of his nature as necessary being. Something might rightly be said to have the property of "being evil," but evil per se is an absence or lack of what ought to be there, and so "being evil" is not a positive property. Anselm leans on this privative theory, and it is a core constituent of Thomas Aquinas' argument that God can be said to cause an evil choice, without His being the cause of evil per se.

causes are "secondary" causes; secondary not in any temporal sense, but in the hierarchical sense that they are dependent upon God for their existence. What caused the cotton to burn? The fire as the secondary cause and God as the primary cause. But God did not "intervene" or "manipulate" either the fire or the cotton. He simply caused them to exist as what they are with all their properties and behavior. So—for purposes of the divine controller argument—we can hypothesize a God who is sustaining everything in being from moment to moment, even you with your choices and subsequent actions. (This does not rule out the possibility of God producing unlooked-for effects: miracles. But in order to counter the "intervener" point all we need is an instance of God causing your choice in a way that does not suggest that you exist independently of God such that He interferes as a limited controller would have to do.)[19] So, while God may be a complete divine controller, He need not be a manipulator or intervener. In the divine controller argument your lack of responsibility for the choice to murder cannot be ascribed to your having been mistreated, used, or even simply manipulated, by the controller.

Could it be that our intuition about the controller is really rooted in the simple fact that we resent conforming to what someone else wants us to do? Surely not. In the course of our lives we do many, many things that other people want us to do. It would be bizarre to suggest that that alone interferes with moral responsibility. Suppose there is an ideal observer who, without causing you to do anything, observes everything you do. And suppose it turns out that everything you do is just what the ideal observer wants you to do. There is no element in this picture to vitiate your moral responsibility. Nor can your lack of responsibility be ascribed to the divine controller knowing ahead of time what you will choose. Divine foreknowledge does not translate into divine, or any other sort, of problematic necessitation of the foreknown. In fact, it is Anselm who succeeds in reconciling true *a se* choices with divine foreknowledge.[20] (I will return to this briefly in Chapter Four.) What is worrisome in the controller scenario is not that someone *wants* us to do something, or that someone might *know* beforehand *that* we will do something. That leaves the worry that someone *makes* us do something.

[19] This is the thesis of Hugh McCann (1995): 582–9. McCann here labels Thomas a "libertarian" in that, as McCann interprets him, Thomas does not hold that the agent's choices are causally necessitated by temporally preceding natural causes, and therefore Thomas denies "determinism," if we mean universal or natural determinism. By the understanding I proposed in the introductory chapter, Thomas's God does determine the choices of created agents.

[20] See Rogers (2008) Chapters 8 and 9.

Taking a cue from Pereboom's progressive, four-case argument, I propose that our intuition about moral responsibility should stay the same as we hypothesize a divine controller whose causal activity is mediated and so comes to look more and more like that at work in a determinist universe. Suppose God sustains the world in being from moment to moment indirectly by immediately causing an angelic intelligence which, in turn, simultaneously with its own causation, causes and sustains everything that is not God or the angel. Suppose God causes a simultaneous causal *chain* of angelic intelligences which cause everything else. Or suppose He causes a super-machine which simultaneously causes everything else. Or a causal chain of super-machines. None of these divinely-caused additions changes the fact that your choice is not "up to you." It is still up to God, but at one remove (or a number of "removes"). None of the new links in the chain adds anything in which your aseity, and hence your moral responsibility, could be grounded.

Suppose, instead of simultaneous causal activity, we hypothesize that the divine controller—now quite distant from the God of classical theism—operates through a temporally successive series of causes, starting before your conception, which brings you and, later, your subsequent choice, into being. Suppose that God should arrange all the necessitating causes for you and your choice to murder within the initial singularity, should light the fuse for the Big Bang, and then, *per impossibile*, should blink out of being. Now the determining chain of natural causes unfolds following the divine plan but without immediate divine guidance up to the point where you choose to murder and commit the murder. *Now* do you deserve to be punished? In that God is the ultimate cause of your choice, the absence of God at the time you come to commit the murder does not inject anything into (or subtract anything from) the situation sufficient to ground your aseity and moral responsibility. (At least that is how the Anselmian sees it.)

And if it is the fact that we are made to do something that conflicts with our having genuine aseity and moral responsibility, then it is difficult to see the relevant difference between some*one* doing the making and some*thing* such as the causes at work in a deterministic universe.[21] There seems to be nothing relevant to distinguish our final hypothetical universe, where God arranges

[21] Perhaps the critic might point to the fact that the controller is one thing while "the causes at work in the deterministic universe" are many and complex (Marius Usher, "Control, Choice, and the Convergence/Divergence Dynamics: A Compatibilistic Probabilistic Theory of Free Will." *The Journal of Philosophy* 103 (2006): 188–213, see 210–13). But it is not clear what sort of relevant distinction could be drawn from this. In any case, the defender of the divine controller argument could construct a scenario in which God introduces all sorts of complex causes in addition to his act of primary causation.

everything and then disappears leaving the chain of causes to unfold, and the deterministic universe without God in its pre-history. If you are not responsible for the choice to murder in the former, then you are not responsible in the latter. If an agent who is divinely controlled is not morally responsible, then an agent whose choices are caused by a deterministic universe is not morally responsible.

A *Tollens* Response?

The compatibilist may agree with that last proposition, but negate the consequent. That is, he may believe that Premise 2 in the divine controller argument is true, but that Premise 1 is false. If a plausible case can be made that Premise 1 is false then the value of the Anselmian project is undermined. Premise 1 just *is* the basic intuition which inspires Anselm and the Anselmian to attempt to construct a viable libertarianism. On the other hand, if the compatibilist's attempt fails, then the basic intuition is strengthened. The compatibilist's *tollens* version of the divine controller argument can be spelled out as follows:

1*. Even if natural causes in a deterministic universe determine your choice, you may nonetheless be morally responsible for it.
2. Natural determination of your choice due to natural causes in a deterministic universe is relevantly similar to divine causal necessitation.
Therefore
3*. Even if God determines your choice, you may nonetheless be morally responsible for it.

The compatibilist may allow that neither the incompatibilist's divine controller argument nor his own *tollens* argument can be shown to have more intuitive power or argumentative support than its rival. And so the discussion over which argument is more plausible ends in a draw. But in that case, at least in terms of the contemporary debate, it could be argued that the compatibilist wins, since it was the incompatibilist who proposed the controller argument in order to show that compatibilism is mistaken.[22]

But this is a bit hasty. There is an asymmetry between the divine controller and the *tollens* arguments.[23] The premise in the divine controller argument—if God causes your choice you are not morally responsible—is an intuitive claim which is

[22] Michael McKenna, "A Hard-Line Reply to Pereboom's Four-Case Manipulation Argument." *Philosophy and Phenomenological Research* 77 (2008): 142–59.

[23] Derk Pereboom (2008) makes a somewhat similar point regarding the standard controller argument, but the case can be made more forcefully in the context of the divine controller argument.

immediate (you see it as soon as you understand the terms), powerful, and widely accepted. The premise in the *tollens* argument—although you are determined you can be morally responsible—certainly cannot lay claim to that sort of intuitive support. To be plausible at all, it must assume a fairly sophisticated form of compatibilism, usually the result of lengthy argument, which argument often begins with reasons for dissatisfaction with libertarianism. The premise in the *tollens* argument does not have the *prima facie* intuitive strength that the premise in the divine controller argument does. That means that when we arrive at the intuitively difficult conclusion that we are morally responsible—we *deserve to be punished or rewarded*—even if our choices are caused by God, our reason to accept the conclusion, rather than rejecting the premise, is comparatively weak.

The compatibilist could argue that if we limit our "intuition pool" to those who are educated about the issues, Premise 1* in the *tollens* argument might have significant appeal based on accepting the truth of the conjunction of two claims; (a) we are free and responsible and (b) indeterminist accounts of choice cannot successfully ground our freedom and responsibility. So, if we are indeed free and responsible, it must be possible for us to be free and responsible on a determinist view.[24] Claim (a) does seem intuitively powerful and widespread. Can (b) make the same boast? Presumably, accepting (b) would be based on having rejected attempts to construct indeterminist accounts of free and responsible action. One might move from there to the conclusion that all past, present, and future attempts at a successful libertarian account—including the present proposal of Anselmian libertarianism—are probably doomed. But if this is what it takes to show that Claim (b) is justified, then accepting it must be preceded by a lot of study. And, of course, in that there is no consensus in the philosophical community now, there is no assurance that future decades will achieve agreement on Claim (b). Thus, unless we draw the circle of the "educated" to ensure the result, it seems wildly improbable that "You are determined yet free and responsible" has the same intuitive force as "You don't deserve to be punished for what God made you do."[25]

[24] Mele, as a "reflective agnostic," offers a tentative suggestion along these lines (2006: 191). John Martin Fischer, too, allows that an agent may be free even if God presets his will by programming his character and motivational states ("Ultimacy and Alternative Possibilities." *Philosophical Studies* 144 (2009): 15–20, at 18).

[25] Were it shown that all libertarian theories must involve some fatal theoretical flaw, the Anselmian would feel free—based on accepting Claim (a)—to grant that libertarian free will exists, but is, and may always remain, theoretically opaque. On the other hand, were she convinced—by irrefutable scientific evidence, divine illumination, or some such—that there is no libertarian freedom, the Anselmian would conclude that we are not robustly free and responsible.

The compatibilist may respond that his version of compatibilism is sufficiently persuasive to justify the *tollens* move. Let us look briefly at a compatibilist theory which explicitly raises the controller argument and makes the *tollens* move to see why the Anselmian remains skeptical. The Anselmian critique here can, I think, be applied with similar effect against other compatibilist views, *mutatis mutandis*. In the introductory chapter I mentioned the complex analysis of autonomy that Alfred Mele offers in his 1995 *Autonomous Agents*.[26] He holds that an agent who meets his proposed criteria may be free and responsible even if determined.

Mele explains that to be psychologically autonomous one must first be an "ideally self-controlled agent." That is (to offer a very sketchy outline) one must meet four criteria: (1) One must have self-control which ranges across all the relevant categories—overt actions, mental actions, intentions, beliefs, etc. etc. (2) One must exercise self-control, not errantly, but in support of decisive better judgments, values, etc. etc. (3) One must exercise self-control whenever one reflectively deems it appropriate, and (4) The exercises of self-control "always succeed in supporting what they are aimed at supporting."[27] In addition to being an ideally self-controlled agent, the psychologically autonomous agent must meet three more conditions, the "compatibilist trio":

1. The agent has no compelled* motivational states, nor any coercively produced motivational states.[28]
2. The agent's beliefs are conducive to informed deliberation about all matters that concern him.
3. The agent is a reliable deliberator.[29]

Note that meeting these criteria enables the autonomous agent to evaluate his values and "shed" them should he find sufficient reason to do so. This is an ability that Mele holds to be central to autonomy.[30]

In responding to the challenge posed by Robert Kane's suggestion of a covert nonconstraining controller (CNC), Mele's first response is to make the move discussed above which distinguishes between natural determinism and divine determinism. He argues that his psychologically autonomous agent could not suffer the sort of control which Kane envisions since the controller must operate through control of the victim's motivational attitudes, as in brainwashing, or through control of the victim's informational attitudes, as in deceit, or through

[26] In Rogers (2012b: 287–8) I also give the example of Lynn Rudder Baker's compatibilist embrace of the *tollens* move.

[27] Alfred Mele, *Autonomous Agents* (Oxford: Oxford University Press, 1995): 121.

[28] The asterisk indicates compulsion not arranged by the agent. Mele (1995): 166.

[29] Mele (1995): 187. [30] Mele (1995): 153, 190.

control of the victim's executive qualities, as in covert conditioning. But the victim of such control would not satisfy the compatibilist trio. Mele grants that the victim of Kane's CNC is not free, but holds that an agent produced by natural causes might be determined, and yet still meet the criteria. The controller argument fails because the causes inherent in the controller scenario are sufficiently different from the causes at work in a deterministic universe to allow us to deny autonomy to the controlled agent in the former, yet still grant it to the determined agent in the latter.[31]

This response to the controller argument does not work when we posit a divine controller. We can change the hypothesis so that God causes you with your choice to murder and causes you in such a way that you meet every one of Mele's criteria. God causes you as a person capable of evaluating and possibly shedding your values. He causes you to be ideally self-controlled. He causes your properly formed motivational states. There is no compulsion. (Compulsion, for Mele, means, in addition to literal, physical force, the sort of state induced by irresistible desire, such as drug addiction.[32] But there is nothing like that here.) God causes the required sort of belief formation and the subsequent beliefs. And He causes the reliable deliberations which lead you to the choice to murder—which He causes as well.

Mele addresses the possibility of something like my divine controller, hypothesizing a creator who creates an adult agent with all of the requisites for autonomy. He concludes, making the *tollens* move, that someone whose choices are caused by (his analogue of) a divine controller in such a way that his criteria for autonomy are met is indeed autonomous.[33] More recently, Mele has introduced a related controller argument, the "zygote" argument. Here Diana, the controller, creates a zygote in a determinist universe, such that it will develop into Ernie who will "*A*"—Diana's desired action—in thirty years. Ernie has all of the attributes and engages in all of the processes Mele takes to satisfy a compatibilist account of sufficient conditions for *A*-ing freely. Analogous to the divine controller argument, the first premise of the zygote argument states that Ernie is not a free and responsible agent. The second premise notes that "there is no significant difference between the way Ernie's zygote comes to exist and the way any normal human zygote comes to exist in a deterministic universe." And the conclusion is that "determinism precludes free action and moral responsibility."[34]

[31] Mele (1995): 187–9. [32] Mele (1995): 136–7.

[33] Mele (1995): 190. His version is most like my case where God arranges everything at the Big Bang.

[34] Mele (2006): 188–9.

Mele holds that, given the point that many of the educated have serious doubts about the possibility of free and responsible action on the part of agents making *non*determined choices, one might intuitively reject the first premise of the zygote argument. In that case, Ernie may be free and responsible even though Diana has created him through his zygote such that he will *A*. Mele says that he is himself agnostic and notes that many factors drive intuitions. For example, if *A* is a praiseworthy action we may be more likely to hold Ernie responsible than if it is a blameworthy one.[35] In response, I grant that our intuitions are inexact and can mislead, but casting the example within the *divine* controller argument simplifies the elements involved and clarifies the situation. The Anselmian insists that there is relevant symmetry involved in God making you do the evil or the good in that in neither case do you choose *a se*. In both cases the decisive causal impetus for the choice is from God and not from you. And given that symmetry, there should be no asymmetry in our views about your deserts. There might be practical reasons why we—limited human beings—should be more eager to praise you for the good you have done than blame you for the evil. But the Anselmian holds that, in the context of the metaphysical debate over free will, an intuition that allows you more credit for your good deed than blame for your bad is a misguided intuition.[36]

Mele's addition of his purportedly "autonomy-making" properties does not address the fundamental Anselmian intuition. If we have qualms about your responsibility when God causes your choice, why should they be alleviated when God causes your deliberating *and then* causes your choice? With the addition of Mele's criteria to the agent in the divine controller example, you are simply doing more things—exercising self-control, deliberating, shedding values, etc.—that God makes you do.[37] If it seemed unjust that you be punished for a murder God made you choose to commit, it seems equally unjust that you be punished

[35] Mele (2006): 193. Mele also notes that our intuitions might be affected by what theory of causation we accept. I noted at the beginning of this chapter that if we adopt a Humean theory, then we cannot mount a plausible controller argument to begin with.

[36] This point counts tellingly against views such as Susan Wolf's which hold that the determined agent who chooses well is responsible, while the determined agent who chooses badly is not (*Freedom within Reason* (New York: Oxford University Press, 1990): 79–81).

[37] Daniel Dennett (1984: 64–5) asks us to consider, rather than the demonic neuroscientist, the "eloquent philosopher who indirectly manipulates a person's brain" through persuasive reasons. If God causes you to choose to murder through causing your mental states and causing the philosopher to present the persuasive reasons such that you are determined, by these factors, to choose to murder, (or to choose to sell all you have and give to the poor) then it seems to me that you are not deserving of blame (or of praise). The reason we are not offended at the thought of "manipulation" by the eloquent philosopher is presumably that we assume that his eloquence does not "induce" our desires, beliefs, and decisions in a way that closes off options and determines us to one choice rather than another. He is not really analogous to the controller in the controller argument.

when God causes you to choose to murder *and* causes you to do all sorts of other things which function as secondary causes, causing you to murder. It seems implausible to hold that you are not responsible for doing X when God directly causes you to do X and only X, but you *are* responsible for doing X when God directly causes you do to X *and* Y *and* Z.

Our question was whether or not contemporary compatibilists could produce an intuitively powerful *tollens* argument against the divine controller argument thus defeating the incompatibilist's attempt to tip the scales of debate in favor of incompatibilism. The challenge was to construct a compatibilism persuasive enough to weaken intuitive resistance to the conclusion that you *are* responsible for your divinely caused choice. But the addition of any number of other elements to the history and background of the choice, if those elements are *divinely caused*, does nothing at all to shake the intuition that if God causes your choice, you are not responsible. Mele grants that divine controller scenarios can be constructed which are sufficiently similar to naturalist causation scenarios such that, *if* you are free and responsible in the naturalist scenario, *then* you are free and responsible in the divine controller scenario. But he does not succeed, by the addition of more divinely-caused elements, in making it plausible that you could be free in the divine controller scenario. Other forms of compatibilism are subject to the same criticism.

Anselm's basic intuition emerges strengthened from this discussion of the divine controller argument. We can safely embrace the claim that if God, or the natural universe, determines all your choices, you are not the morally responsible, metaphysically valuable being that Anselm understands you to be. The task, then, is to try to construct a theory which defends the aseity of human choices. In making this attempt Anselm proposes a view which is new to the current discussion, though similar enough to contemporary libertarian theories to connect with them in interesting ways. In the rest of this work I spell out an Anselmian libertarianism based on Anselm's work, and then attempt to defend the theory against the standard criticisms that are leveled against libertarianism, and also to address a contemporary problem raised against the thought that we "create" ourselves by our free choices. But before I embark on the main work of the book, I would like to add one more point in favor of accepting Anselm's approach.

Moral Laxness and a Wager

I noted in the introductory chapter that, within the Anselmian project of self-creation, how *you* view yourself and your choices is extremely important.

Suppose that you are not convinced by the divine controller argument. You think that, at best, the libertarian and the compatibilist have arrived at an intuitive draw, and that this applies not just to the divine controller argument, but to the state of affairs in the free will debate in general. (For purposes of this section understand "compatibilist" to mean "determinist compatibilist," someone who believes that determinism—universal or natural or theist—is true, but that human agents are nonetheless to be judged free and responsible. One could be a compatibilist and hold that free will is compatible with both the truth and the falsity of determinism.) You see advantages and disadvantages to both libertarianism and compatibilism. (You take hard determinism to be right out, since you find it non-negotiable that you are more or less responsible.) And you are uncomfortable not adopting one view or the other. Why might you feel uncomfortable? Well, you cannot escape making choices. And perhaps you are sympathetic to the Anselmian thought that part of your job, as you go about your business in the world, is to try to become a better person through your choices. If you are interested in the project of self-creation, then you are a self-reflective person. So you must have some view of yourself and your choices. Sitting alone in your room, doing philosophy, you may be able to maintain an agnostic stance, but as you live your life, when you actually debate between options or look back on what you have chosen, it will be difficult to sit the fence between libertarianism and compatibilism.[38] These are very different theories of what constitutes a choice and, if you are thinking hard about your choices, it will be a struggle not to assume or adopt one or the other. If you hold that science and philosophy have, to date, failed to decide the question between libertarianism and compatibilism, then you might engage in some Pascalian-style wagering and ask which view is better to adopt; better insofar as the project of self-creation is concerned.[39]

Let us suppose that you are confronted with a morally significant choice. For example, though you see that it would be wrong, you are sorely tempted to be unfair in connection with an evaluation for promotion and tenure of a colleague who has been obnoxious to you and who, adding insult to injury, has published more in better venues than you have. But you are also committed to trying to be

[38] Balaguer (2010: 69) suggests that, since science has not proved it either way regarding free will, "All we can do at the present time is shrug and say, 'Who knows?'." But that won't do in that we are forced to choose and act and think about our choices and actions.

[39] Studies have documented numerous benefits from believing oneself to be free, many of which are related to the present question of moral laxness. See David G. Myers, "Determined and Free." In Baer (2008): 32–43, at 38–40; Azim F. Shariff, Johnathan Schooler, Kathleen D. Vohs, "The Hazards of Claiming to Have Solved the Hard Problem of Free Will." In Baer (2008): 181–204. Eddy Nahmias (2010: 345) notes that the media suggests that determinism provides a universal excuse for all behavior.

an honest person. You are torn, trying to decide what to do. Grant that it is hard to decide to do the right thing. Now, as you are deliberating before your choice, if you are a compatibilist, you might add some "weight" to your temptation to be dishonest, saying to yourself, "Well, whichever choice I make, I really can't choose otherwise. So I might as well stop struggling to do the honest thing. I might as well relax and do what I want to do. I'll try to sabotage that obnoxious hot shot!" And suppose you do so. Afterwards you might feel ashamed and ruminate on the choice you made. But remembering that you are a compatibilist, rather than feeling regret and trying to make amends, you might well succumb to the temptation to dismiss your uncomfortable feelings about your choice on the grounds that you were determined to choose what you chose and you couldn't have chosen otherwise in any case. My suggestion here is not that the compatibilist fails to deliberate. He does debate about what he ought to do. The thought is rather that, assuming one sometimes has to work hard at making the morally good choice, one might factor in one's compatibilism as a reason for not making such an effort. If you fail to make the effort, it happens that you *couldn't* have made the effort.

This point—that determinism, even in its compatibilist mode, is likely to encourage moral laxness—goes back at least to the fifth century. When St. Augustine introduced compatibilism to the main stream of Christian philosophy, holding that all of your choices are caused by God, it created quite a stir. "If whatever you, in fact, do is whatever God has made you do, why bother to try so hard to be good?!?" asked Augustine's opponents rhetorically. The same argument was made in the ninth century when Gottschalk revived Augustine's compatibilism with what was apparently too much clarity.[40] But just because it is an old criticism doesn't mean it is not a telling one. Psychologically, doesn't it seem that if you think of yourself and your choices from a compatibilist perspective you are less likely to work at making the harder moral choices and more likely to make excuses for yourself? Undoubtedly many compatibilists, and even hard determinists, are paragons of virtue. The claim is just that thinking that you are determined could lead to moral laxness.

[40] Rogers (2008): 112, 113; 125–6. In the theist universe, on the hypothesis that absolutely every event is caused by God, there is the extra problem that whatever *anyone else* does is exactly what God is causing them to do. And, at least on the classical theism of Augustine and Aquinas, God must be causing them to do what they do in order to produce some greater good. Why, then, should we work so hard to try to prevent other people from doing evil when, if we fail, it is because God wants the evil deeds to produce an even greater good? See my *Perfect Being Theology* (Edinburgh: Edinburgh University Press, 2000): 147–9.

The compatibilist may respond that the libertarian, too, faces the problem of moral laxness. Doesn't the libertarian believe that, in a given choice, either non-determined option is compatible with every fact about the past and the laws of nature? And doesn't this show that the choice for this over that—ending up in *this* possible world over *that* possible world—is just a matter of luck? Confronting your choice, if you are a libertarian, you can say, "Well, since whichever choice I make, it's just a matter of a mental coin flip, I might as well relax and do what I want to do. I'll try to sabotage that obnoxious hot shot!" And afterwards you might dismiss your choice as unlucky, but not something that you need to regret or be ashamed of.

There is an asymmetry here, though. Actual compatibilists (as I am understanding the term here) do not deny that your choice was determined. Actual libertarians *do* deny that your choice was lucky. Libertarians argue, for example, that luck seems to be what happens to you, whereas making choices is something you do. In Chapters Seven and Eight I address the luck problem, arguing that those who find it damning for libertarianism have mischaracterized the act of choice, that the possible worlds version of the luck problem does not add anything to the original characterization of the problem, and, perhaps most important, that the originators of the luck problem got the relationship between choice and character and responsibility temporally backwards. The compatibilist who insists that the problem of moral laxness applies equally to the libertarian must suppose that libertarianism *entails* that choices are lucky in something like the way compatibilism (as I am understanding it here) entails determinism.[41] But that means he is supposing that none of the currently available attempts to respond to the luck problem is or could be successful and that future attempts will likewise fail. This supposition is rendered all the more dubious in that, at least in the contemporary literature, the debate seems to be largely driven by intuitions. In order to set up the wager all that needs to be granted is that it is intellectually respectable (not obviously contradictory) for you to hold a libertarian theory and deny that choices are a matter of luck. In the wager below "libertarianism" will mean the theory that the agent bears ultimate responsibility for his non-determined choices *and* that the luck problem can be managed such that the agent need not believe that choices are just a matter of a mental coin toss. Someone who understands libertarianism this way will not be tempted, on the basis of holding the view, towards moral laxness.

[41] Mele (2006: 70) at one point defines "luck" such that it seems to be an entailment of the definition of libertarianism. That would just beg the question against the libertarian, but, of course, Mele has a great deal more to say to try to motivate the intuition that libertarianism involves responsibility-denying luck.

The compatibilist might suggest that the libertarian is tempted towards a different moral fault than laziness, and that is pride. Couldn't there be something unseemly in the libertarian insistence that *I* am the master of my fate and *I* am the captain of my soul? The original Pelagians, against whom Augustine refined his compatibilist views, could certainly be accused of this failing. They were upper-class Romans who wanted to rise above hoi polloi by their own efforts at moral purity. Among Augustine's motivations was a consistent anti-elitism which the modern reader is likely to find welcome, even if she does not share his compatibilism.[42] But it will become clear in Chapter Three that Anselm's own libertarian theory includes a safeguard against an unwholesome pride. The safeguard arises out of the parsimony of Anselm's theory, in which the aseity of the free choice does not require the introduction of any new entities or causes. All goods, in Anselm's system, come ultimately from God. All the free agent does, in choosing rightly, is to pursue a God-given, good motivation when he could pursue an inappropriate (God-given) motivation. So the agent can take only a very little credit for his good choices and should not get a swelled head on the basis of simply having clung to the God-given good. The same is true even of an earth-bound Anselmian libertarianism that does not explicitly include God as the source of all. If it is nature that supplies the options, pursuing the good over the bad is a modest contribution on the part of the free agent. Still, it is *some* contribution, and the agent who is trying to stay on the right side of the universe (whether or not it is the product of a good God) should be happy to take responsibility for her humble efforts. So libertarianism, at least in its Anselmian version, need not tempt anyone to pride. On the other hand the criticism that compatibilism is likely to lead to moral laxness has not lost its strength even though it has been around so long.

So here's the wager on the hypothesis that believing that your choices are determined will tempt you to moral laxness, while believing in libertarianism will not. We'll give you a point (+1) if you reap a benefit, and we'll take away a point (–1) if you suffer a harm. (Benefits and harms include matters having to do with the Anselmian project of self-creation, so being tempted to moral laxness is a harm.) Suppose you believe in compatibilism. If compatibilism is true, you have the advantage of believing the truth (+1), but the disadvantage of being tempted to moral laxness (–1), so your score is 0. If compatibilism is false, you have the disadvantage of believing what is false (–1) and the disadvantage of being tempted to moral laxness (–1), so your score is –2.

[42] Rogers (2008): 129.

Now suppose you believe in libertarianism. If libertarianism is true, you have the advantage of believing the truth (+1), and the advantage of not being tempted to moral laxness (+1), so your score is +2. If libertarianism is false, you have the disadvantage of believing what is false (−1) and the advantage of not being tempted to moral laxness (+1), so your score is 0. Do the math. Libertarianism is a better bet!

And, in any case, if you are a libertarian and libertarianism is false we cannot ascribe to you any *ultimate* responsibility for your failure to believe the truth. You couldn't have done any differently. On the other hand, if compatibilism is false, you may well be ultimately responsible for your failure to believe the truth. You are better off in terms of the project of self-creation if you believe that you are free in a libertarian sense.

Or think of it this way. Suppose you have a dear friend who is struggling against gluttony or alcoholism. He's been good about his diet, or hasn't had a drink, for eight months. But now he desperately wants that cupcake or that one little drink, and is debating within himself whether or not to give in to temptation. You fear that if he goes for the cupcake or the little drink, that might start him on a slide back into active (as opposed to recovering) gluttony or alcoholism. You want him to succeed in resisting temptation, and you are thinking about what to say to help him. You might give him the libertarian pep talk: "You can do it! You really don't have to eat (or drink) that! It's up to you to decide!" Or you might say, "You'll do whatever you are determinined to do. Whatever you end up doing was the only thing you could *actually* do."[43] Expressing the compatibilist position is probably worse than not saying anything to your struggling friend. Again, in terms of self-creation, it is better for your friend to believe that in some absolute way, it is open to him, and up to him, to resist temptation.[44]

Of course this doesn't prove libertarianism. All it shows is that you are probably better off believing libertarianism than compatibilism, in terms of your life's work of creating yourself through your choices.[45] But if you hold that science and philosophy have left the scales pretty evenly balanced between

[43] There are, of course, all sorts of compatibilist analyses of "you could do otherwise," but these all involve qualifications that are consistent with the determinist point that you could not actually do otherwise.

[44] Actual Alcoholics Anonymous advice includes a very strong component of asking for help from a "higher power." But even the asking involves a choice, and even if one has (or believes one has) received divine aid, one has to choose to embrace it.

[45] It is often said, even by those who deny libertarian free will, that it is better for us to embrace the illusion of free will. See for example, Saul Smilansky (2010: 187–201).

the two, then here's a reason to lean towards libertarianism. This wager comes into play only if you hold the evidential scales to be evenly balanced between libertarianism and compatibilism. For myself, I wholeheartedly agree with Anselm's basic intuition, and I find the divine controller argument decisive against compatibilism.

2

Anselmian Libertarianism

Background and *Voluntates*

> Clearly *voluntas* can be spoken of equivocally in three different ways. The instrument for willing is certainly one, the *affectio* of that instrument another, and the use of that same instrument, yet another.
>
> *De Concordia* 3.11

Introduction

In this chapter I begin to set out Anselm's theory by looking at how he understands the relationship of the will as a faculty to the motivating elements in a choice and the actual desiring which can come to constitute a choice. These are the three meanings of the term *voluntas* that Anselm distinguishes. But before I embark on this project, more background is in order. Throughout this work I will be comparing and contrasting Anselm's theory to some representative contemporary event-causal and agent-causal views, and so it is helpful to say a word about each and briefly note the standard problems that are raised against each. Then I will set out Anselm's motivation for constructing his theory. This is part of my methodology of emphasizing the underlying worldview in which his analysis of free will arises. I think this helps to clarify the position and to explain Anselmian intuitions. Anselm's motivation proves to be interestingly similar to that of Robert Kane in that both hope to present libertarian theories which do not invoke any new causes which are unique to human choosing. This point sets the stage for my discussion, in Chapter Three, of how Anselm achieves a *parsimonious* agent-causation.

Then I set out and defend a couple of preliminary methodological points: First, I will follow Anselm in discussing only choices with moral significance. Second, I will stick with the example Anselm uses, the fall of Satan. Although Satan has not appeared on stage much in the contemporary debates, there are several reasons why it is helpful to retain the diabolic example for purposes of setting out the Anselmian theory. Finally I will discuss the careful, analytic distinctions that Anselm makes between three meanings of "will," *voluntas*. This will establish the elements out of which Anselm constructs his libertarian view, and so the three need to be on the table first in order, in Chapter Three, to work through Anselm's theory of how *a se* choice is possible.

Event-causation and Agent-causation

The Anselmian holds that only a libertarian account of human freedom can include the aseity which is required to support the possibility of self-creation and the traditional, and still common, thought that human beings have a unique kind of dignity. Some forms of compatibilism may allow for some qualified ascription of moral responsibility, praise and blame, and reactive attitudes associated with freedom. But, as Anselm sees it, only libertarianism can ground the special metaphysical stature which human beings have been thought to possess. It is a worthy task, then, to attempt to set out the strongest libertarian theory available.

In the contemporary literature there are a number of different versions of libertarianism. Some libertarians argue that it is best to view a choice as an uncaused, basic action.[1] But this approach, as Randolph Clarke points out, does not seem to provide an adequate account of the sort of control the agent should be thought to exercise over the choice, nor of how the agent's reasons can play a role in explaining how the choice comes to be.[2] Those who find these criticisms telling propose causal accounts of libertarian freedom which, to date, have fallen into two categories, the event-causal account (event-causation) and the agent-causal account (agent-causation). The former holds that events within the agent non-deterministically produce a free choice. The latter argues that it is the agent himself who non-deterministically causes the choice.

Anselm is properly located in the agent-causal camp. He is an inheritor of the standard Aristotelian approach: When it comes to efficient causality—a taking action which brings some thing or event into being—what is most properly considered a cause is a substance, a particular, concrete thing which exercises a power to produce some effect.[3] As I noted in the introductory chapter, the

[1] Two important proponents of this view are Carl Ginet, *On Action* (Cambridge: Cambridge University Press, 1990) and Hugh McCann, *The Works of Agency: On Human Action, Will and Freedom* (Ithaca: Cornell University Press, 1998).

[2] Randolph Clarke, "Libertarian Views: Critical Survey of Noncausal and Event-causal Accounts of Free Agency." In *The Oxford Handbook of Free Will* edited by Robert Kane (Oxford: Oxford University Press, 2002): 356–85, at 356–61.

[3] *Philosophical Fragments* in *Memorials of St. Anselm* edited by R. W. Southern and F. S. Schmitt, O.S.B., (London: Oxford University Press, 1969): 344–9. Anselm here lists various meanings of the term *facere* which map onto various modes of "doing." He remarks that insofar as we are speaking

contemporary Anselmian need not be embarrassed by the assumption of substance causation in that the view, especially as related to agency, is currently making a comeback.[4] It lies outside the scope of the present work to weigh in on the question of substance causation in general. What is important here is that, beyond simply accepting the standard causal theory of his day, Anselm explicitly agrees with the agent-causalists that the locus of responsibility is the choice, which is up to the agent alone. An *a se* choice is not caused by anything except the agent. It is not caused—as event-causation would have it—by preceding factors within the agent, even probabilistically.

But Anselm's position is interestingly different from some of the more prominent agent-causal theories such as those recently proposed by Clarke and O'Connor, which take agent-causation to be a special and uniquely human sort of causal power. When I speak of agent-causal theories I will usually be referring to these sorts of theories. There are others. Helen Steward has recently attempted a very naturalized understanding of agency wherein the minimal requirements for an indeterminist agency can be met by lower animals, perhaps as low on the chain of being as spiders.[5] In that Steward very explicitly eschews attempting an analysis of responsible, rational, agency, her explication of "agency indeterminism" is very distant from the Anselmian's concerns. Thus it is not useful to present it as a major foil against which to set out the Anselmian theory. However, some points of comparison, such as the treatment of lower animals discussed in the next chapter, will be noted in passing.

Anselm's approach agrees with a common criticism of standard agent-causation made by compatibilists, by event-causalists like Robert Kane, and also by Helen Steward. The criticism holds that it is ad hoc to introduce into one's theory special sorts of causation which are unique to human freedom. Anselm argues that the agent produces the choice, but not through any new form of causal power. The agent chooses through exercising the motive power of preceding desires, such that Anselm's descriptions of choices and their histories seem, in some respects, reminiscent of Kane's descriptions. And Anselm is clear that the sort of motive power he has in mind is not uniquely human. Lower

about efficient causes, then only the sort of cause that acts to bring something about really *does* what it is said to do. If a killer should kill someone with a sword, for example, the killer is an efficient cause.

[4] See, for example, E. J. Lowe, *Personal Agency* (Oxford: Oxford University Press, 2008); Helen Steward, *A Metaphysics for Freedom* (Oxford: Oxford University Press, 2012). Anselm's theory of personal agency is quite different from either Lowe's or Steward's, though.

[5] Steward (2012): 108.

animals have wills which are moved through these same sorts of powers.[6] For purposes of comparison a word more about event-causation and agent-causation is in order before setting out the preliminaries to Anselm's own position, which has affinities to both.

Robert Kane, the most influential event-causalist, is motivated to produce a libertarian theory that does not invoke any entities that are somehow unique to the theory. In this, he hopes to remove a stumbling block to the scientifically-minded. He writes that:

> In the attempt to formulate an incompatibilist or libertarian account of free agency ... we shall not appeal to categories or kinds of entities (substances, properties, relations, events, states, etc.) that are *not also needed by nonlibertarian* (compatibilist or determinist) *accounts of free agency satisfying the plurality conditions* [the agent faces open options]. The only difference allowed between libertarian and nonlibertarian accounts is the difference one might expect—that some of the events or processes involved in libertarian free agency will be indeterminate or undetermined events or processes.[7]

Kane holds that the nondetermined choice can be explained in terms of preceding "efforts" involving motives, or reasons, or intentions, which are events or states of affairs within the agent. In a paradigmatic case of free choice, an agent engages in "efforts of will" to realize mutually incompatible purposes. This constitutes an indeterministic process which terminates in the production of the choice.[8] Kane continues the theme of a theory with which the scientist can be comfortable, suggesting, tentatively and as a "working hypothesis" that:

> The indeterminate efforts of will ... are complex chaotic processes in the brain, involving neural networks that are globally sensitive to quantum indeterminacies at the neuronal level. Persons experience these complex processes phenomenologically as "efforts of will" they are making ... The efforts are provoked by the competing motives and conflicts within the wills of the persons ... These conflicts create tensions that are reflected in appropriate regions of the brain by movement further from thermodynamic equilibrium, which increases the sensitivity to micro indeterminacies at the neuronal level and magnifies the indeterminacies throughout the complex macro process which, taken as a whole, is the agent's effort of will.[9]

Given our present state of knowledge, I fear that this attempt to connect the choice of an agent with goings on in the brain—especially sub-atomic goings on!—will not really advance the discussion. If we take Kane's description to involve the "butterfly effect"—a tiny event here is the beginning of a causal

[6] So Anselm agrees with Steward that there are important similarities between the agency of humans and that of lower animals, though his approach is very different from hers.

[7] Robert Kane (1996): 116. [8] Kane (1996): 128–31. [9] Kane (1996): 130.

chain which ends in a large event there—then it looks like Kane is saying that it is the indeterminate "bouncing" of a particle (or some particles) this way rather than that which is the initiating cause producing a choice. If the particle bounces *this* way then the agent will choose X, if it bounces *that* way, then the agent will choose Y. If this is the picture he intends, then it appears that the choice is causally necessitated by the non-determined behavior of a particle. And unless the bouncing of the particle can somehow be identified with the conscious choice of the agent himself, this is a form of determinism, by my definition.[10] I take it that Kane does not really intend this "butterfly effect" picture. For one who wants to connect or identify conscious choice with brain activity, it might be best to admit that, as things stand at present, we cannot do better than note that there is indeterminacy throughout the universe, and that in complex systems such as the weather and the human (or perhaps the animal) brain events may occur which are not determined.[11]

Kane emphasizes the fact that the agent's eventually choosing this over that can be explained as probabilistically caused by his *motives* or *reasons* for choosing this.[12] Kane notes that many contemporary philosophers defending compatibilism insist that we must say that agents act for reasons, and that these reasons play some causal role. He writes, "I think libertarians can accept this assumption so long as the causal connections between reasons and actions are allowed to be probabilistic, as well as deterministic."[13] Kane's picture, then, is one of an indeterminate process in which an agent is debating between competing reasons or motives, which process ends in a choice that is probabilistically caused by one of the reasons or motives.

There are a number of problems that can be raised against Kane's view, but the standard critique from a libertarian perspective claims that this event-causal theory does not provide the agent with the sort of control required for robust freedom which could ground moral responsibility. Randolph Clarke, spelling out an agent-causalist view, argues that Kane's event-causation, in holding that the choice is nondeterministically caused by its immediate causal antecedents, differs

[10] Jason Turner raises this point against Kane, "The Incompatibility of Free Will and Naturalism." *Australasian Journal of Philosophy* 87 (2009): 565–87.

[11] Steward, for example, identifies decisions with brain happenings, but, wisely in my view, does not try to get specific about how this could work.

[12] Kane (1996): 136–7; see also "Libertarianism." In *Four Views on Free Will* edited by John Martin Fischer, Robert Kane, Derk Pereboom, and Manuel Vargas (Oxford: Blackwell Publishing, 2007): 5–43.

[13] Kane (1996): 136. Mark Balaguer presents roughly the same picture of a libertarian free choice. He writes that the choice "flowed out of his reasons and intentions in a nondeterministically causal way." *Free Will as an Open Scientific Problem* (Cambridge, MA: MIT Press, 2010): 80.

from a compatibilist account only by entailing indeterminism. Comparing a compatibilist agent and a Kane agent, Clarke argues that the former is not relevantly different from the latter except that in the case of the Kane agent, the choice, "resulted from an indeterminate effort and that, until the decision occurred, there was a chance that the agent would instead make an alternative decision right then." But in either case the choice is caused, deterministically or probabilistically, by factors for which the agent does not bear responsibility. Clarke concludes that if the compatibilist agent is not responsible, then neither is the Kane agent.[14]

What is missing from the event-causal account, says Clarke, is the positing of "further positive powers to causally influence which of the alternatives left open by previous events will be made actual."[15] Clarke suggests an "integrated agent-causation": In addition to the agent's motives and reasons, we should posit the agent himself as a substance cause of his choice.[16] Timothy O'Conner, an agent-causalist of a somewhat different stripe, notes that, in addition to event-causes, "According to the agency theorist, there is another species of the causal genus, involving the characteristic activity of purposive free agents."[17]

Critics accuse standard agent-causation of introducing a new brand of causation which they hold to be ad hoc and which, they argue, cannot be plausibly developed. Clarke himself concludes on a pessimistic note regarding agent-causation, due to problems with the underlying theory of substance causation.[18] Kane writes that strategies, including agent-causation theories which involve introducing "extra factors,"

> do not solve the problems about indeterminism they are supposed to solve and create further mysteries of their own. Moreover, "extra factor" strategies have tended to reinforce the widespread criticism that libertarian notions of free will requiring indeterminism are mysterious and have no place in the modern scientific picture of the world.[19]

I do not entirely endorse Kane's assessement. As I will note in Chapter Seven, agent-causalist views do fare rather better against the luck problem than does Kane's event-causation. Moreover, there really is no developed "scientific picture" of how agency, or even just consciousness, actually works. Add to that the thought that history teaches that at least large swaths of "the modern scientific picture of the world" will be superseded by future scientific pictures

[14] Randolph Clarke, *Libertarian Accounts of Free Will* (Oxford: Oxford University Press, 2003): 107. Timothy O'Connor offers a similar criticism in *Persons and Causes* (Oxford: Oxford University Press, 2000): 36–42.

[15] Clarke (2003): 108. [16] Clarke (2003): 133–49. [17] O'Connor (2000): 71–2.

[18] Clarke (2003): 185–212. Steward responds to Clarke specifically (2012): 220–4.

[19] Kane (2007): 24–5; see also Kane (1996): 120–3.

of the world.[20] The science of the brain is in its infancy, and so it seems rather restrictive to require that a theory of free will be a good fit with whatever the current consensus is among those who study the brain.[21]

On the other side of the coin, it seems defeatist and contrary to fact to insist, as Immanuel Kant does, that free will is a creature of the noumenal world, in principle outside of our experience and beyond the reach of science. We are able to experience our choices to some extent, and it is far too early in the game to decide that science will not be able to tell us a great deal about how our free will works. Modern agent-causalists do not go the Kantian route. Instead they try to show that the "extra factor" involved in agent-causation is not as mysterious as its critics make it out to be. In the present work I do not pass judgment on recent efforts by agent-causalists to answer their critics. My aim is to show that Anselm's libertarianism succeeds in ascribing more control to the agent than does Kane's event-causal theory, while at the same time (though it does assume a standard substance-causal approach) explicitly avoiding the introduction of any new, *sui generis*, or "mysterious" extra factors unique to human willing. If Anselm's move is successful, then parsimony would suggest that Anselmian libertarianism is preferable to standard agent-causal theories.

Motivation for Anselmian Libertarianism

Interestingly, Anselm and Kane are motivated to elaborate libertarian analyses of free choice for analogous reasons. Both insist that compatibilist theories do not provide the free will that we *want*.[22] Both insist that only a libertarian theory

[20] In that physicists look at one aspect of the world, biologists at another, psychologists at another, etc. etc., it seems unlikely that there is *a* scientific picture which could be relevant to the free will question. Sometimes philosophers and the man on the street suppose that there is such a monolithic, well-developed, and cohesive "picture." And some scientists, when they are on television or dabbling in philosophy, seem to support that thought. But in fact there does not seem to be such a thing. Moreover, the conclusion that we have *now* achieved the apex of Science such that our present picture, if such a thing exists, has delivered immutable truth and so should be the arbiter of our metaphysics, is deeply debatable. In every age since Thales got the disciplines of philosophy and science rolling there have been those who insisted upon roughly that same conclusion—"modern" science sets the parameters within which metaphysics must proceed—but that conclusion has consistently been proven mistaken in the past.

[21] A glance at the literature from experimental psychologists studying the connection between free will and brain activity shows them to be rather unsophisticated when it comes to understanding the free will issue in the critical way that philosophy requires. See Mele (2009).

[22] At the beginning of his *De casu diaboli* (DCD) Anselm dismisses the, then popular, Augustinian compatibilist theory to explain why Satan fell (see Rogers (2008: 73–6)). The first half of Kane's *The Significance of Free Will* is devoted to showing why compatibilism does not allow us enough free will.

involves a sort of free choice robust enough to ground what we are looking for, the *ultimate* responsibility of the agent.[23] And they agree on one more thing, although the motives behind this agreement arise from very different background presuppositions. They both hope to construct plausible theories that do not introduce any new things or causes unique to the process of free choice. Kane, as noted above, finds it unfortunate that so many libertarian theories hypothesize mysterious types of agent-causation. He hopes to develop an analysis which can support our freedom, but which does not generate any tension with a scientific approach by adding new entities or principles beyond those discussed in the sciences.[24]

Anselm, of course, has no concern to produce a theory that will "have a place" in Kane's late twentieth-century scientific picture of the world—if there is such a thing. But he does confront an analogous puzzle regarding the multiplication of entities. Anselm's task is to find room for human aseity and ultimate human moral responsibility in the universe of classical theism.[25] In this universe all the concrete objects and their powers and positive properties are brought into being and sustained from moment to moment by an omnipotent God. But, as the divine controller argument in Chapter One holds, if God causes our choices we do not have *a se* freedom. Now the question for Anselm is: If God is the cause of all that exists, how can our choices possibly *not* be caused by God?

Like Kane, Anselm hopes to present an analysis of our ultimate responsibility which does not depend upon positing any new entity or power invoked specially to allow for our freedom. In Anselm's case this is because the introduction of something new with ontological status would not solve the problem of what, in a choice, could be up to us, since any real "thing" must come from God.[26] (I use "thing" in a somewhat restricted way to be discussed at length in Chapter Four.) The question for both Kane and Anselm is this: Is it possible to allow for the ultimate responsibility of human agents without introducing new entities and principles beyond what we already take to exist, having been produced either by God or by natural causes? (In Anselm's universe it is better to say "by God *and* by natural causes." The former is the "primary" cause as that which produces and sustains the existence of things, the latter constitutes "secondary" causation. Classical theism holds that one of the goods of our universe is that created things have and express causal powers.)[27] So the motivation behind Anselm's theory is

[23] "Aseity" for Anselm (Rogers (2008: 76–8); "Ultimate Responsibility" for Kane (Kane (1996: 33–7).

[24] Kane (1996): 115–17. [25] Rogers (2008): 16–29. [26] Rogers (2008): 108–23.

[27] Thomas Aquinas, who had confronted the theory of occasionalism through the work of Algazali, argues that a created object with its own causal power is an intrinsically better sort of

similar to Kane's. But his theory is interestingly different. For Anselm it is the agent, not events within the agent, that causes the choice. But there is more at issue than just the basic underlying Aristotelian assumption of overall substance causation. Anselm's theory requires both that the free agent chooses with aseity and yet that the choice does not depend upon, or introduce, or in any way involve, elements not made by God and hence "new" to the universe.

Anselmian Preliminaries

Anselm grants that there are many different sorts of freedom and many venues for choice, and explains that the choices he is concerned about are choices with moral significance, choices for which the agent might be appropriately praised or blamed.[28] He takes it that a creature capable of moral responsibility can engage in the project of self-creation. This sort of creature is a much more valuable thing, it enjoys a much more elevated metaphysical status as an *imago dei*, than other creatures which cannot make morally responsible choices.[29] As I noted in the introductory chapter, I will follow Anselm in limiting the scope of the discussion to morally significant choices. These do seem to be the choices that are of most concern in the free will literature. But note that, in Anselm's tradition, choices with moral content cover a wider range than the contemporary philosopher might suppose. Anselm, a good eudaemonist, does not make a hard and fast distinction between moral choices and prudential choices.[30] Your over-arching life goal—the aim of your self-creation—is to flourish as a human agent, and your choice of school to attend, career, vacation spot, etc., likely has some value content that will affect your character and the overall kind of person you are. So the focus is not as narrow as it might seem at first. And some of what is said of moral choices might apply to other sorts of choices as well.

What Anselm is definitely not concerned with is choices of indifference. That is, the Buridan's ass sort of choice where the options are not just equally desirable, but they are the same in all relevantly interesting respects. If your choice is between chocolate and pistachio ice cream, and you like them equally and are equally in the mood for each, each is readily available, etc., etc., then you might just as well flip a coin. Your choice here will have no subsequent impact on your character. The choice of indifference may well be one species of free choice, but it is not of interest to us here.

thing than one without and a better representation of the power of God (ST I, Q.105, art. 5). See also my *Perfect Being Theology* (Edinburgh: Edinburgh University Press, 2000): 113–18.

[28] *De Veritate* (DV) 12. [29] DCD 16. [30] Rogers (2008): 60–72.

Anselm sets out his analysis in three dialogues which he asked to be always bound together in order; *De Veritate, De libertate arbitrii,* and *De casu diaboli* (*On Truth, On Freedom of Choice,* and *On the Fall of the Devil*). Later he wrote a work attempting to reconcile free will with divine foreknowledge, predestination, and grace, *De concordia praescientiae et praedestinationis et gratiae dei cum libero arbitrio* (*On The Harmony of God's Foreknowledge, Predestination, and Grace with Free Will*), which elaborates on some of his earlier views. His is, admittedly, a very basic outline of a libertarian theory. We can hardly fault him for this, in that he is probably the first to attempt a systematic description of the processes involved in a libertarian free choice. In terms of the case I want to make in this chapter and the next chapter—that Anselmian libertarianism has affinities with various contemporary libertarian theories but is a different and attractive view—the simplicity is a good thing. It allows the structure, the "bones" of a choice to be seen clearly. The rest of the book will put some meat on those bones to produce a somewhat more developed theory. And if this first pass at Anselmian libertarianism proves plausible, it should set the stage for subsequent work aimed at a richer and more complete theory.

Anselm takes as his main example of a free and responsible choice, the fall of Satan. Angelic doings have played little role in the contemporary free will discussion, and yet, for our purposes in this chapter, there are good reasons to stick with this example in spelling out Anselm's theory.[31] First, it is an idealized instance of choice which allows us to focus on the structure of the event without worrying about the plethora of present and historical circumstances which surround an actual human choice. (Chapter Five argues against those who insist that it is the history of a choice that is all important, not the structure.) For purposes of setting out Anselm's theory consideration of such circumstances would be extraneous and might muddy the issue by invoking irrelevant intuitions. Various issues arising from messy human reality will be addressed later in the book.

Another reason that the Satan example is helpful is that it focuses attention on a choice, rather than on an overt deed. This is exactly where Anselm and the Anselmian want to focus. The claim will be that responsibility for an overt deed derives from having freely chosen to do the deed. Overt deeds can be labeled "freely done" only insofar as they result from freely made choices. Moreover, in

[31] Steward (2012: 32) understands agency as involving bodily movements, but does not present any argument, beyond a citation to a work that briefly mentions the difficulties with classical substance dualism. Satan, we may suppose, is not embodied, and so the problems with dualism do not arise.

that our concern is with self-creation, it is important to note that many of our most morally significant choices may not involve overtly doing any deed at all. I may make choices about the kind of person I want to be or the sort of character I want to have. So, for example, if I choose to forgive rather than nurse my anger, that choice need not issue in any overt deed at all. It may, but it need not. Perhaps the subject of my forgiveness is dead, and so I will not be called upon to engage in overt action towards them at all. Furthermore, how we choose to evaluate our overt deeds after we have done them may be more important in terms of character-creation than the doing of the deeds. (I will discuss this in Chapter Eight.)

Further, the Satan example, in that it involves God and a catastrophic choice, underscores the necessity of aseity defended in Chapter One. And the Satanic example also includes a ready-made contrast class, appeal to which makes it very vivid why responsibility requires that it not be one's past that determines all of one's choices. Anselm's concern is to explain the fall of Satan, as against the contrastive choice of those angels who did not fall, although, in all the essentials, their desires and motivations were, *ex hypothesi*, the same as Satan's. So, although it introduces an unfamiliar cast of characters, I will stick with Anselm's example.

At the beginning of *De casi diaboli* Anselm reviews a standard explanation for Satan's fall. This is Augustine's position, although Anselm does not mention Augustine's name.[32] The theory is that, because they are brought into being from nothing, all of the angels suffer from a tendency to abandon the good appropriate to them by pursuing lesser goods. God gives some the extra help to resist that tendency. These hold fast to the original good and can now be referred to as the "good angels." The ones to whom God fails to give the extra help inevitably succumb to temptation and can now be referred to as the "bad angels." We have to say this, Augustine opines, since otherwise we would have to hold that the angels who chose well had made themselves better on their own. And that's absurd![33]

But this theory just cannot be right, says Anselm, for then the ultimate responsibility for the bad choice would be up to God who made the angels with the inherent tendency and then failed to help some.[34] Anselm insists that we cannot explain the choice of Satan, or of the good angels who made a different choice, through any factor preceding the original choice, since absolutely everything about the angels, their motivations, their character—everything!—comes from God. If the cause of the choice is ultimately traceable to God, then the created agent is not responsible.[35] Aseity is the foundational property allowing

[32] Rogers (2008): 30–52. [33] Rogers (2008): 45–6. [34] DCD 2.
[35] This point is made throughout *De casu diaboli* and especially in Chapter 27.

for freedom, responsibility, self-creation, and the elevated metaphysical status that the Anselmian wants to ascribe to the created agent.

Three Meanings of *Voluntas*

Although I am more concerned with the philosophical viability of the Anselmian view than with historical interpretation, it seems well to stick pretty close to Anselm as he builds his theory. He is careful and subtle *and* inventive, and I think spelling things out in his terms will be the most effective approach. Thus I begin to develop Anselmian libertarianism with Anselm's own analysis of meanings of the term *voluntas*. That is the best way to begin to get the elements of choice on the table. Anselm is an analytic thinker who is constantly concerned that our use of language may lead us astray. That being the case, he undertakes to point out that we might attribute to the term *voluntas* three different meanings which need to be distinguished.

First, there is *voluntas* considered as an instrument, a faculty of or in the soul.[36] I take it that this is quite an old-fashioned view. But Anselm does not suggest that the will is some discrete "thing." It is a power of the soul, and, for those who are unsympathetic to souls, it can be thought of simply as the agent's ability or power to want and to choose things. (Those who will not admit abilities and powers will find this aspect of the Anselmian approach misguided.) He sometimes says that the agent "turns" the will as instrument.[37] This might suggest that, preceding some act of will, there must be an earlier act on the part of the agent by which he turns his will. And that introduces a regress. I think we can read Anselm as intending that, in engaging in an act of will, the agent exercises his power, without supposing that there must be some previous act as an impetus. This seems a natural understanding of his view that the will is a power in the soul by which the soul wills what it wills. And, insofar as Anselm does not subscribe to Cartesian dualism, but rather to Augustinian dualism in which the human being is a *unity* composed of soul and body, we can read this as the same as saying that the will is that power by which the *agent* wills what he wills.

But the will, considered as an instrument and in isolation, does not choose, or "move" as Anselm puts it. Within the will as instrument, there is a second sort of *voluntates*. These are the *aptitudines* of the will as instrument. *Aptitudo* is

[36] The clearest and most developed discussion of the three *voluntates* is in *De concordia praescientiae et praedestinationis et gratiae dei cum libero arbitrio* (DC) 3.11. All of the explanation concerning the three meanings of *voluntas* is taken from that chapter.

[37] *De libertate arbitrii* (DLA) 7.

apparently an uncommon Latin term.[38] But we can tell from the context and from its root in the Latin *apto*, that it means something related to the Anglicized word "aptitude"; a sort of power aimed at, or suitable for, doing something. In this case what the "aptitudes" in question do is "move" the will towards the objects which, one way or another, are suitable for the agent. Anselm calls these motivating aptitudes *affectiones*. We can think of the *affectiones* as desires, or, more broadly, just motivating factors, but his understanding of these *affectiones* does not map perfectly onto any concept in the contemporary literature. For one thing, Anselm is careful to note that an *affectio* may be occurrent or not. He holds that, for instance, it is proper to say that we always desire health—we have an *affectio* to be healthy—whether or not we are thinking of it at the moment. The *affectio* is always there, and then, as soon as we think about being healthy, we occurrently desire it.

A second interesting feature is that, as the term "aptitude" suggests, there is a teleology to these *affectiones*. They are *for* achieving something. My impression is that, in the contemporary free will literature, desires and motivating factors are often treated as neutral phenomena. Yes, the agent has subjective goals or purposes, but these are not necessarily connected with objective and overarching goods for the agent.[39] Anselmian *affectiones*, on the contrary, are aimed at goods appropriate to the nature of the creature. They represent a proper goal-orientedness on the part of the will of the agent. (Things can go wrong, of course, as we will see below.) Anselm's own thinking is richly teleological, rooted as it is in classical theism and owing a robust, if indirect, debt to Aristotle. In setting out the Anselmian position, I will speak of desires and motivating factors, and perhaps the reader who is unsympathetic to a teleological approach can safely bracket it insofar as appreciating the basic Anselmian picture of the workings of the will is concerned. But the fact remains that Anselmian libertarianism is rooted in a metaphysics that finds *objective purpose* lurking everywhere, and this must color the whole theory.[40]

[38] It does not appear in *Lewis and Short*, the standard Latin dictionary.

[39] The evolutionary goals of surviving and reproducing are, of course, very important in the Christian context. But surviving and reproducing are set within a larger and richer teleological picture than evolution in a universe of atoms and the void can provide. If it's just atoms and the void, then you going about your business, including having children, and the human race's continuing a bit longer before the sun turns into a red giant or the universe dies a heat death, does not have the ultimate importance all of this business has on the Anselmian picture in which you and your children are everlasting.

[40] Kane ("Libertarianism." *Philosophical Studies* 144 (2009): 35–44, at 35) sometimes writes that the theory he defends allows us as free agents to have "the power to be ultimate originators of at least some of our own ends or purposes." Anselm would not express the situation exactly this way. In general terms our ends or purposes are objective and given by God and our natures. There may be all

The third meaning of *voluntas* refers to the actual use (*usus*) of the will. This is when the agent—due to the *affectiones*—is thinking of what he wants and occurrently desiring it. It is significant that Anselm chooses to express this meaning of *voluntas* as "use," in that, at least ordinarily, when we *use* something we are actively *doing* something. The sort of occurrent desiring he has in mind seems to be an active exercise of the will. His examples involve activities, with the suggestion that they are to be engaged in immediately—"Now I want to read," "Now I want to write."[41] In the context though, it looks like an occurrent desire to be healthy, even if it does not immediately translate into a trip to the gym, would constitute a use of the will. And that use of the will should be thought of as an agent's *doing* something, even if all he is doing is wanting to be healthy.

Anselm does not make as many distinctions here as one might wish. He does not, for example, consider the phenomenon of what we might call "idle wishing" of the "I-wish-I-could-flap-my-arms-and-fly" sort. I suspect that in his picture of how the will works this sort of idle wishing is off of the spectrum of the related meanings of *voluntas*. It would be a sort of entertaining imagining, but not the sort of wanting or desiring that constitutes the will as use. I think it is safe to say that *voluntas* as use may sometimes be simply an occurrent desire, or it may be something more, what we might term a "volition" or an "intention," that is an action-guiding plan.[42] In that we are translating the term *voluntas*, "volition" might be the better term, but "intention" is, I believe, more common in the literature, so I will use it to express that mental state which goes beyond "I want to..." to "I'm going to...". And this "I'm going to..." is not a mere belief about the future, but rather incorporates the "oomph" to get the deed done. (There is a significant qualifier to this point about "getting the deed done." "Intention" does not map exactly onto "*a se* choice." This is another reason why using the term "volition" might be confusing. All this will be discussed in Chapter Three in describing the *a se* choice.) As Anselm explains the will, both desiring and intending should be thought of as actions—in a very broad sense of action—on the part of the agent. Or, if "action" seems too robust a term to apply to a desire,

sorts of good ways of pursuing these general purposes—someone gifted with acute critical skills might debate between being a philosopher or an analyst for the FBI, for example—but even here, we are not the originators of our interest in these options.

[41] This looks rather like E. J. Lowe's use of the term "volition," in discussing Locke's theory of the will (*Locke on Human Understanding* (London: Routledge, 1995): 119–41), but there are clear differences. For one, Lowe (and perhaps Locke?) seems to assume that a volition must be part of the causal process of moving a part of the body, and Anselm, of course, understands the use of the will more broadly in that it may or may not have to do with overt deeds.

[42] See my "Anselm on the Ontological Status of Choice." *International Philosophical Quarterly* 52 (2012a): 183–98, at 195.

say that it is at least a "doing." In the contemporary free will literature having a desire is often described as, or assumed to be, simply passive, but Anselm doesn't see it that way.[43] We do not create our *affectiones*, and they're part of ourselves whether or not we are aware of them, but when we "use" them, we are doing something.

Anselm does not discuss the point that, phenomenologically, we might feel that a desire has "come over us," or that we are "in the grip" of a desire, or that we passively find ourselves in certain "appetitive states." I might, for example, notice that I'm beginning to feel sleepy. This does seem like a desire to sleep, but it does not feel like something I'm *doing*. There is no reason for the Anselmian to deny such a phenomenon. On the Anselmian account this "feeling" may grow out of the unconsciously possessed *affectio*. (There are, of course, all sorts of physiological events occurring in the body of the person who begins to notice that she is sleepy. But that fact in no way undermines the thought that there is a general, teleological *affectio* for well-being, which can be manifested as noticing that one is becoming sleepy and then as a more active desire to sleep.) What is important is that, as we come to consciously entertain the desire, we are, if only in a minimal way, doing something. The importance of this point will be clear in Chapter Three when the *a se* choice is described. Before choosing, the Anselmian agent engages in dual "efforts" (to use Robert Kane's terminology)[44] to realize incommensurable objects of desire, and this very active way of describing the prelude to the choice is crucial to the overall picture.

It is also important to note that, as Anselm describes the process, there is a continuum between a simple occurrent desire and an intention. (Anselm's discussion of the use of the will is all too brief, but thankfully we can turn to Alfred Mele's series of helpful distinctions concerning intentions. We may distinguish between a standing intention and an occurrent intention, and between a distal—aimed at a future action—intention and a proximal—aimed at an immediate action—intention.)[45] For our purposes "intention" will refer mainly to an occurrent, proximal intention. Anselm's thought is that an occurrent desire may evolve organically into an intention at the point when the agent's simply wanting X becomes his embracing an action-guiding plan to do or obtain X. A desire may also fail to so evolve. The claim here is not that desires *cause* intentions, but rather that the intention just *is* the desire continued past a certain

[43] Thomas Pink, for example, says that having a desire is passive. "Freedom and Action Without Causation." In *The Oxford Handbook of Free Will: Second Edition* edited by Robert Kane (Oxford: Oxford University Press, 2011): 349–65, at 357.

[44] Kane (1996): 126–8. Lowe describes a volition as a "trying" (1995: 120).

[45] Mele (2009a): 3–10, with references to earlier works.

point, such that it is fully embraced.[46] This is why the Anselmian will not see having a desire and forming an intention as two radically different sorts of events, one passive and one active. This thought, that desiring is an act which can evolve into intending, is one of the key theses that allows for the parsimony Anselm is hoping to achieve with his theory. With the three *voluntates* distinguished and explained, the elements are in place to set out Anselmian libertarianism.

[46] In saying that the intention is the desire, I do not mean to "reduce" the intention to something lacking in executive power. Rather the thought is that the agent's desiring something can constitute a changing process in which it reaches a point of "settledness" such that the agent now fully plans to put the "wanting" into practice. The Anselmian, thus, rejects a hard and fast distinction between reductive and non-reductive theories regarding intention, in which an intention is either simply a desire/belief pair or is not a desire at all. For this distinction, see Alfred Mele, "Introduction." In *The Philosophy of Action* edited by Alfred Mele (Oxford: Oxford University Press, 1997): 1–26, at 19.

3

Anselmian Libertarianism

A Parsimonious Agent-causation

Let us say..., even though it is not customary, that to persevere in willing is to "per-will" (*pervelle*).... So therefore I say that the devil, who received the will and the ability to receive perseverance, and the will and the ability to persevere, nonetheless did not receive perseverance and did not persevere because he did not per-will it.

De casu diaboli 3

Introduction

With the three meanings of *voluntas* on the table, we can begin to develop the Anselmian description of an *a se* choice. It is helpful at the outset to discuss Anselm's understanding of animal willing. Highlighting the (very profound and interesting!) similarities that Anselm sees between animal and human willing, as well as the significant dissimilarities, will help make Anselm's parsimonious agent-causation clearer. One crucial difference, as I will go on to explain, is that rational agents can be torn between morally significant options—and these alternative possibilities set the stage for *a se* choice. But Anselm is insistent that the *a se* choice does not require any special causal powers unique to human choosing. Thus, Anselm's libertarianism is properly labeled agent-causal, but it is parsimonious in that, unlike standard agent-causal theories, it does not invoke any *sui generis* powers. I explain the mechanics of this parsimonious libertarianism, then conclude the chapter by responding briefly to the suggestion—which standard agent-causalists might make—that Anselm's theory is *too* parsimonious.

Animal Willing

Let us take S to stand for some rational, created, agent. (I choose "S" because it also stands for Satan, the main character in Anselm's analysis of free choice.)

In Anselm's universe, S and the factors that originally motivate S, come from God. As outlined in the previous chapter, the *affectiones* provide the motive power, the "wind in the sails" of our acts of will. Anselm understands the agent's having certain desires as a kind of cause, and, of course, assumes that this cause involves some force or efficacy. In the interest of allowing the created agent to make himself better on his own God organizes the *affectiones*, the various motivating factors, to produce alternative possibilities.[1] One could adopt the basic structure of Anselmian libertarianism without embracing the sorts of alternatives that Anselm suggests, but that there are alternatives is non-negotiable. As Anselm spells it out, there are two categories of good which S might pursue, benefit and justice. Benefits, *simpliciter*, are those things (in a broad sense of "thing") which S believes will make him happy. And, given certain crucial qualifications, they are the sorts of things that *would* make him happy. (God-given desires are teleological, aimed at the actualization of the nature of the creature.) We are always motivated by a desire for benefit. We would not consciously pursue anything at all if we did not consider it to be of benefit.[2]

Anselm holds that rational beings are not the only agents. Like humans and angels, lower animals like dogs and horses have *voluntates*, wills as faculties or instruments, possessing *affectiones*.[3] A standard criticism leveled at libertarians is that they "de-naturalize" the human will, seeing it as something not of a piece with the rest of the universe. Helen Steward writes, "In an era which had yet fully to embrace the idea that human nature is continuous with that of the rest of the animal kingdom, perhaps it was unsurprising that human agency should be singled out in this way for special consideration."[4] I am not sure which "era" Steward has in mind, but I suspect it must be post-Cartesian. Anselm certainly does see human beings as special and different from lower animals—as I'm sure we all do—but that does not mean that in terms of will and agency he supposes human nature to be discontinuous with the rest of the animal kingdom. On the contrary, it is interesting and useful to see how very *continuous* the human and animal will are in his view. He gives animals a great deal of credit, although it should be noted that in terms of his explication of human freedom, nothing really hangs on whether or not his views on animal agency are correct.[5]

[1] Augustine had found that suggestion absurd, so Anselm is carving out a new and radical position here.

[2] Rogers (2008): 66–72.

[3] For the horse see *De Veritate* (*On Truth*, DV) 12; for the dog see *De libertati arbitrii* (*On Free Choice*, DLA) 13.

[4] Steward (2012): 2.

[5] Hans-Johann Glock argues that recent research supports the position that animals can be properly described as acting "for reasons," as self-aware, and as capable of deliberating; "Animal

Anselm assumes, *pace* Descartes, that lower animals (hereafter "animals") are the subjects of inner conscious experience which is something like much of the inner experience of human animals. (Presumably he supposes that animals can form beliefs or something like them. One traditional distinction which medieval philosophers saw between the thought processes of lower animals and that of human beings is that human reason involves theoretical and scientific thinking. A horse can register the thought that this patch of grass is desirable, but it cannot think *about* grass in a way that would answer the question, "What is it?") When Anselm is building his definitions of "justice" and of "free will" step by step, it is all but the last steps which can apply to both animals and human beings. Part of the definition of "justice" is "preserving rightness of will." Animals have wills, and, like us, they are motivated by the *affectio* for benefits. Moreover, animals can "preserve rightness of will." This "rightness of will" means willing appropriately; willing what enables the animal to flourish or—it's the same thing—willing as God would have the animal will. One can imagine a sick or injured animal failing to will rightly, but a healthy animal wills rightly as a matter of natural necessity. And because they will as a matter of natural necessity they cannot will *justly*. "Justice" is not simply rightness of will, but rightness of will *preserved for its own sake*. And it is this "preserving for its own sake" that lies beyond the powers of the animal, since one must be rational to be able to "step back" and examine one's own desires.[6] In that "free will" is the ability to preserve justice, animals cannot have free will by Anselm's definition.[7] But their powers of willing are continuous with ours until that "stepping back" point. In terms of willing, the horse and the dog can will rightly and thus be a good horse and a good dog. What they cannot do is step back, examine their own desires, and freely identify with, and act upon, the appropriate ones. They cannot *choose that they should be* a good horse and dog.[8] Anselm even underscores how *very* continuous the wills of humans and animals are when he notes that, if a rational agent (he is talking about Satan)

Agency." In *A Companion to the Philosophy of Action* edited by Timothy O'Connor and Constantine Sandis (Wiley-Blackwell, 2010): 384–92. As he explains these abilities, they do not approach the sort of rational agency which Anselm ascribes to human agents.

[6] DV 12. Augustine, too, suggests a similarity and continuum which ends where human beings can "step back." He notes that we share with animals our five senses and then a sixth, "inner" sense, which governs the five, unifying the different sorts of information and telling us what is to be pursued and what avoided. We part company with our non-rational fellow creatures at the point where the rational human being is able to "look within" and consider his own perceptual powers (*De libero arbitrio* 2.3–4).

[7] DLA 5.

[8] I discuss the interesting similarity to Frankfurt's criteria for personhood in Rogers (2008: 60–2, 66–72). See Harry Frankfurt, "Freedom of the Will and the Concept of a Person," *Journal of Philosophy* 68 (1971): 5–20.

were given only the *affectio* for benefit, and not for justice, and if he found himself unable to have "greater goods," then he would will lesser goods of the sort that "delight irrational animals." And, of course, there's not a thing wrong with animals pursuing these lesser goods, since they do so by nature. Nor would S be blameworthy for willing as animals will in this case, since he would not be free and able to make *a se* choices.[9]

It will help to clarify the *a se* choice, by comparison, if we take a moment to consider an animal will at work. Along Anselmian lines we can explain the horse strolling out to the pasture by saying that it wanted to eat grass, and so formed an intention* to go graze. (The contemporary literature almost always takes intentions to be rationally formed, so the asterisk indicates the equine equivalent of a rationally formed plan—it is the condition of willing that actually gets the legs moving.)[10] We might add an earlier stage—a point at which the horse passively experienced a feeling of hunger "growing on him." Phenomenologically, that often seems to be the case with hunger. But—and this, too, seems right phenomenologically—that passive feeling can quickly elide into a desire which becomes an active wanting on the part of the agent. (At least that is how Anselm sees it, as I discussed in Chapter Two.) And the agent, in this case the equine agent, can then move from simply wanting to intending*.

Anselm himself—having no interest in whatever brain events may constitute or accompany the process of choice—does not address the question, but there is nothing in the Anselmian schema which precludes the horse forming an intention* through a physically non-determined process. A subscriber to event-causation might attribute to our horse open options along the lines suggested by Robert Kane. Presumably equine brains are quite complex, and so, if one allows that sub-atomic indeterminism may produce macro-indeterminism in very complex systems, it is not obvious why the sort of description that Kane gives for the physical processes which probabilistically cause a choice could not be going on in the horse. Perhaps he is Buridan's horse, and waffles between two equally appealing tufts of grass. And perhaps there are subatomic indeterminacies in his brain such that an indeterminate process will probabilistically cause him to opt for the tuft on the left, although, had the subatomic indeterminacies fallen out differently, he would have opted for the tuft on the right. Or perhaps the horse was thirsty as well as hungry, and—again due to an indeterminist

[9] DCD 13.

[10] O'Connor (2000: 55) says that intentions require intelligence as well as purposiveness. Anselm clearly holds that animals—at least higher animals like dogs and horses—can act purposively, and thus it seems he would attribute to them the sort of willing that involves the continuum of desire becoming something very like an intention.

process in his brain—could have pursued his desire to drink, rather than his desire to eat. So perhaps the horse strolling over to the tuft on the left is—from the perspective of secondary causation—caused probabilistically not deterministically. To my knowledge, Kane does not address the question of choice in lower animals. But if the non-determined process which he finds so crucial for freedom could be roughly the same for lower animals and for humans, this counts against his position from the Anselmian perspective. Though animal willing is not totally unlike human willing, human free choice is something special. The dog and the horse—even on Anselm's optimistic assessment of animal agency—do not choose with aseity.

With regard to the horse's forming the intention*, if it is the behavior of the subatomic particles that causally produces the intention*, then the intention* is "determined" by my definition. (I do not insist that this captures the relationship which Kane intends to describe between the physical process and the conscious choice.) Once that subatomic particle has "bounced" this way or that, the die is cast and the horse as conscious agent must form the intention* which he forms. If there is some viable way to *identify* the subatomic bounce with the conscious formation of the intention*, then perhaps we should not say that the particle bounce *causes* the formation of the intention*, but rather is part of the process which *is* the formation of the intention*. Still, this undetermined, probabilistic causation is not enough to render Buridan's horse free or responsible. And what is lacking is not just rationality, though rationality is certainly a necessary condition for a morally significant choice. Because the horse lacks rationality, it cannot step back and evaluate its options, which means it cannot confront morally significant options, and so it cannot choose with aseity.

On Anselm's substance-causal view, in that the horse has a will and *affectiones* which are powers in its equine soul (in Anselm's day everything animate had an *anima*), it would be proper here to say that it is not preceding events in the horse, but rather it is *the horse* that causes the intention* as it exercises its power to will. This is faintly reminiscent of Helen Steward's view that animals, as organic wholes, exercise top-down causation on their parts and are non-determined agents. Her picture seems to be that animals "settle" how the open future will go in all sorts of ways—building the web here rather than there, starting into the pasture on the left hoof, when starting on the right was equally a possibility.[11]

[11] Steward (2012). Her book is more an explanation and conceptual defense, than an attempt to provide evidence that her view represents what is the case. She argues that to act is to non-deterministically "settle," and it would be contrary to our ordinary view of things to deny that actions happen.

Were science to come to the conclusion that animals are indeed the non-determined agents that Steward proposes, this would, from an Anselmian perspective, advance the libertarian cause in rather the way that the scientific community's allowing sub-atomic indeterminacy did. That is, once science admitted indeterminacy anywhere at all, universal determinism ceased to be "the scientific view." If lower animals act indeterminately, then there is indeterminacy at the macro level, and not just in the weather, but in animals! And that would be a striking blow against those who insist that there cannot be indeterminism at a macro level. But it would be a far step from the horse's option to start on the left or the right to anything resembling the sort of agency that Anselm is interested in.

The Steward horse, like the Kane horse, may confront open options, but without moral significance, the options are uninteresting. As Steward describes animal agency it applies to the minor details of building the web here rather than there, or starting on the left hoof. Although her animals are non-determined agents, she does not dispute that they pursue their goals by nature. Anselm himself would probably hold that even if animals are agents which, in the trivial details, act indeterministically from the perspective of natural laws, it is nevertheless the case that God causes all these physically non-determined actions. If our horse starts out on the left, that is because God is sustaining him in being, with all his properties, and his actions, including his left-originating stroll. Divine determinism of animal—as opposed to human—behavior is not a problem. When animals preserve rightness of will, by willing the benefits they are made to will, it is by natural necessity. The well-functioning, web-building spider cannot opt out of building a web, be it here or there. The well-functioning horse cannot opt out of grazing, though perhaps he might start on his left or his right. When animals don't pursue the proper goods, there is nothing blameworthy about the failure; something has gone wrong with the organism due to the network of secondary causes at work in the world. On the Anselmian account, in terms of the objects they pursue, animal wills are naturally necessitated at the level of secondary causation. And in all of their actions, both the trivial (left or right) and the interesting (grazing), they are divinely necessitated at the level of primary causation. And there is nothing to worry about here since animals are not praiseworthy or blameworthy the way rational agents can be. Nor are they as impressive as we are. We are rational and we can make ourselves better on our own. We can help in our own creation, by choosing with aseity.

Alternatives and the Torn Condition

Anselm has it that what distinguishes the rational agent from the lower animal—in addition to being able to think theoretically—is that the rational agent is able to desire and pursue the good of justice *from itself* or *on its own*. One is just when one deliberately, and on one's own, pursues only those benefits—a subset within the larger set of basic (perceived) benefits—which one recognizes as morally appropriate, *because* they are morally appropriate.[12] This will set up the morally significant alternative possibilities that are required for aseity. In setting out Anselmian libertarianism, it is not crucial that the theory entail exactly the morally significant options which Anselm proposes. Anselm sets things up this way because he will not have a perfectly good God endowing the created agent with motivations that are intrinsically bad. So he cannot say that there are bad desires, or desires for the bad, over against the good desires or desires for the good. One should think of Anselm's proposed motivational structure through a sort of Venn diagram. There's the big circle which is the set of all the agent's desires for benefits. These are real benefits for good things with nothing intrinsically wrong about them. And then there is the space carved out for the subset of desires for benefits appropriate to this agent, in these circumstances, at this time, etc., etc. So, for example, the desire for food is an intrinsically good human desire, but perhaps one should not pursue the desire for this large slice of chocolate cake right now. For simplicity's sake I will stick with Anselm's proposal in which the morally significant options are between what we can call "mere" benefit and justice, but the mechanics of a free choice could be the same on other views of what constitutes morally better and worse choices.

While the Anselmian description of a free choice maps comfortably onto other analyses of what a morally significant choice entails, *that* there is an objective moral or value order is non-negotiable. It is an inherent part of the theory that there *be* better and worse choices. The theory holds that choices produce character and what is ultimately of interest is the kind of person you are in terms of your character. The Anselmian assumption is that there is an objective right and wrong, virtue and vice, virtuous or vicious character. If this assumption is false, then the motivation for adopting Anselmian libertarianism is undermined.[13]

[12] Anselm notes that a rational agent might deliberately pursue the appropriate objects out of a desire for some lesser good, such as to gain wealth, rather than out of a desire to be just (DLA 13).

[13] In that Robert Kane is very concerned about *moral* responsibility, he seems to allow objective values, and hence, contrary to his stated goal in developing his event-causal theory, he seems to allow some "entities" that are not part of the universe of at least some hard determinists and compatibilists and that are not the subject matter of any of the sciences. Objective values lie outside of the "scientific" view of the world if that means the world as described by the various sciences. Even the study of "moral" education in psychology and sociology studies how children come to acquire the mores of their societies, and seems to have little if anything to do with an objective morality.

A further question arises concerning what it takes to "recognize" some benefit as morally appropriate or not. In the Introduction to this work I enunciated the principle that an inclusive theory is, *ceteris paribus*, preferable to one that would eliminate large swaths of normal human adults from the company of those with moral responsibility and elevated status. Thus, I propose that, regarding the criteria for a minimal, basic responsibility the bar for what constitutes recognition of a morally better and worse should not be set very high. In Anselm's own example Satan knows with perfect clarity—he is an angelic intellect, after all, and God has told him what not to do in no uncertain terms—that he should not choose what he ends up choosing. He knows, as well as anyone knows anything, that it is the wicked choice. (What he doesn't know is what the consequences will be for himself. On Anselm's theory, if Satan had known that choosing wickedly would plunge him into abject misery, he would have been psychologically incapable of choosing it, and then he would have "preserved rightness" out of fear, which is not being just.)[14] It would set the bar far too high to insist that to be a free agent requires having Satan's degree of firm belief, and with a divine pedigree behind it!

It would take us too far afield to try to determine the minimum sufficient conditions for relevant recognition, so suffice it to say that at least some, perhaps dim, recognition is required on the part of the agent that there is some morally better or worse distinction between the options he confronts. It is probably best not to insist that the agent himself must be clear on the distinction, so long as he appreciates the relevant facts. I say this in order to take account of instances of agency such as exemplified by Huckleberry Finn. Huck (a fictional, but not unrealistic character) is famously torn between, on the one hand, turning runaway slave, Jim, in to the authorities and, on the other, keeping mum and helping Jim continue his escape. Huck, in some sense, judges turning Jim in to be the morally correct thing to do. It is what society teaches. Mark Twain and we, the readers, see that it is the wrong thing to do. Huck stays mum and is the better person for it. He recognized the key morally relevant factor; what would happen to Jim if he were turned in to the authorities. Even though Huck was not thinking about the situation with the clarity Twain conveys to the reader, the Anselmian should say—assuming that the other conditions for an *a se* choice apply—that he made a free choice, opted for the morally better one, and is praiseworthy. Suppose he *had* understood the situation better. Suppose he saw that helping Jim was the right thing to do, but had some desire, based on fear for himself, to turn Jim in. In that case his choice to help Jim might be even *more* praiseworthy,

[14] Rogers (2008): 95–6.

in that it is more deliberate. For purposes of the discussion here, it is best to allow that the recognition of the moral nature of the options required for a moral choice may fall well below the clarity possessed by Satan in Anselm's discussion.

We always choose what we believe will be of benefit to us, but we can also choose justice, which means choosing to pursue only those benefits which are appropriate for us to pursue. Anselm here foreshadows Frankfurt's account of "personhood."[15] One must be able to step back from one's immediate desires, assess them, and identify with some over others. The horse can pursue benefits appropriate to its equine nature, and so it can desire and intend* rightly. But it cannot survey its desires, consider what it ought to desire in order to be a good horse, and then identify with the appropriate equine desires.

Anselm is clear that the desire for justice is—to adopt Frankfurt's terminology—a second-order desire; that is, it is a desire about what basic desires for benefit should be embraced and followed and what inappropriate desires should be restrained. And this introduces a possible criticism. I take it that most of us working on free will agree that rational human agents are able to exercise *some* control over what desires they have and *some* control over the extent to which those desires influence them. But Anselm here seems to suggest that we have a great deal of control over which desires we will entertain and allow to influence us. Those who hold that we have only a very minimal control will find this approach uncongenial. How can someone be held responsible for what desires they have, if they exercise only a very little control over them?

Here is one area where one's underlying worldview may color one's analysis of free will. Those outside the Christian tradition may well be split on the question of whether or not human agents exercise much control over their motivational states. The Christian, on the other hand, is rather stuck with the position that we can, do, and should, exercise a great deal of control over our desires and inner attitudes. We are told to love God and our neighbor. And Jesus, in His uncompromising way, makes it clear that it's not just that we should *act* like we love God and neighbor. What goes on "in the heart" is as telling as what one does overtly. Killing is wrong, but so is hatred. Adultery is wrong, but so is lust.[16] Anselm's assumption that we have a robust ability to embrace some of our desires and reject others is a non-negotiable pillar of Christian thought. Outside of introspection and anecdote, it may be difficult to *prove* Anselm's point to the doubter.

[15] Rogers (2008): 60–6.

[16] Matthew 5:20–32. Thomas Nagle in his "Moral Luck" in *Mortal Questions* (New York: Cambridge University Press, 1979) includes having virtues and vices as a matter of moral luck and holds that we have little, if any, control over them. He provides only intuitive evidence for his claim.

But perhaps the sort of Pascalian wagering that we engaged in in Chapter One is in order. We do all have desires which influence our behavior. Shouldn't we *try* to exercise control over them? So—if science leaves it an open question—isn't it better to act upon the assumption that we can often succeed in nourishing our better motivations and starving our worse? At the very least, Anselm's thought that we have the ability, at the second order, to evaluate and embrace or reject certain first-order desires has a noble pedigree and contemporary defenders. It is the critic's burden to show that there is something inherently implausible in the hierarchical motivational structure Anselm proposes.

God could engineer an agent's, say S's, motivational structure so that S would desire only the morally appropriate benefits. Presumably—as suggested in the divine controller argument in Chapter One—God could cause S to engage in the second-order assessment of his basic (first-order) desires and cause him to identify with those which are appropriate. Frankfurt holds that the harmony between first- and second-order desires is sufficient for freedom. Anselm disagrees. If the whole process is caused by God, S would will rightly as a matter of divinely imposed necessity, and so he could not be free or responsible or just.[17] (A choice could be necessary and unfree, as divinely imposed, even if it is the result of a non-determined process. That could be the case if the choice is caused by God, and God's acts are not determined. It could also be the case if a choice is caused by the non-determined behavior of subatomic particles, unless that behavior can be identified with the act of choice of the conscious agent.) For S to be praiseworthy S must be able to survey his basic desires, recognize which are morally appropriate, and then *choose on his own* to pursue the appropriate desires. For an agent that is created and dependent, that requires confronting open options, where it is absolutely up to the agent which option he pursues. The stage is set for a free choice only when S desires both an inappropriate benefit and simultaneously wants to be just. (Again, one might assess the moral distinction at work differently without any impact on the basic Anselmian theory of free will.) That means S is torn, struggling to pursue different, mutually exclusive, morally interesting desires. Call this the "torn condition," TC.[18]

[17] DCD 13–14.

[18] The critic may argue that insisting that responsibility is rooted in TC is too strong a requirement. For example, can't we be responsible for failing to do something when we just forgot? The Anselmian responds that we are not responsible for forgetting "through no fault of our own." But if we made the wrong *a se* choice earlier, and that led to the forgetting, then we are responsible. If I should have written it on my calendar and chose not to bother to do so, I am at fault. This introduces the tracing thesis; we can be responsible for choices and actions—or in this case omissions—that do not involve TC if those choices etc. are the product of past *a se* choices. The tracing thesis is the subject of Chapter Nine.

TC sounds rather like Kane's description of an agent making a self-forming choice involving plural voluntary control.[19] And it *is* rather similar. But, besides the fundamental distinction between the event-causal and the agent-causal view, at least two important differences should be noted. First, Kane often speaks of the agent involved in conflicting "efforts" or "willings." An advantage of this sort of language is that it emphasizes the thought that the agent is engaged in an active internal struggle. And that seems right regarding the sort of choice that is the paradigmatic instance of a self-forming choice and the sort of choice Anselm is interested in. But Kane's description invites a criticism: Phenomenologically, while we are all familiar with the experience of struggling between choosing to do what we ought, as opposed to what we would rather do at the moment, it does not seem that a choice involves *actually trying* to achieve two conflicting goals. Indeed, if a person is rational, he can't really be trying or willing (in some robust sense of those terms) to simultaneously achieve objects he recognizes as mutually exclusive, can he?[20] Kane responds to the criticism by suggesting that the agent need not be aware of the dual efforts. They may be going on below the level of consciousness as parallel processing in the brain.[21]

The torn condition as Anselm describes it might suggest this criticism, but the Anselmian does not respond the way Kane does. Yes, the agent *actively* wants or desires objects which he recognizes as mutually exclusive, but desiring or wanting conflicting objects is an all too common experience.[22] The "efforts" language is appropriate, as long as it indicates that conflicted struggle which precedes the actually trying to achieve a goal. But TC, according to the Anselmian, and contrary to Kane's suggestion, is very decidedly a conscious experience, whatever brain activity may accompany or instantiate it.

Kane uses the "tryings" and "willings" language in his attempt to respond to the point that indeterminism in the process which precedes the choice introduces an element of luck which undermines responsibility. But this points to another important difference between how Anselm and Kane understand the torn condition. Kane locates the relevant indeterminism in the torn condition which precedes the choice, whereas Anselm locates it at the moment of choice.[23] This

[19] Kane (1996): 107–15.

[20] See Laura Ekstrom, "Free Will, Chance, and Mystery." *Philosophical Studies* 113 (2003): 153–80, at 163–4; Clarke (2003): 88–9.

[21] Kane (2007): 34.

[22] John Lemos argues that Kane could and should have stuck with the "wanting" language that he sometimes uses rather than discussing the "tryings" and "willings"; "Wanting, Willing, Trying and Kane's Theory of Free Will." *Dialectica* 65 (2011): 31–48. Kane might allow this suggestion, as long as we can understand "wanting" as a species of "doing."

[23] Dennet suggests that it would be insane to put the indeterminacy at the point of choice, and then goes on to describe such a situation as having decided on the best course of action, but then

will bear interestingly on the Anselmian response to the luck problem, which will be discussed briefly in this chapter, and then more fully in Chapters Seven and Eight.

An agent cannot engage in an *a se* choice without first being in TC. At some point—unless S dies, the world ends, etc.[24]—S makes the choice to pursue one option over the other. S's choice looks like this: At time (t)1 S is in TC regarding whether to opt for A or B, where A and B represent mutually exclusive courses of action to be pursued, and where one is morally preferable to the other. (Let A represent the just option and B the inappropriate benefit.) Then, at t2, which occurs after t1 and constitutes the moment of termination of TC, S opts for B over A. At t2 and thereafter, it is true to say that, although S chose B, S could have chosen A.

In analyzing a choice, Anselm is motivated by the theological puzzle involved in his commitment to God as the *creator omnium*, and so he carefully distinguishes the location of the causal "force," the motivating power involved in the choice, from the location of the indeterminism. He takes the causal force to be a thing with ontological status. The competing desires for mere benefit and for justice are what provide the motive power, and they come from God.[25] The choice consists in S desiring the two options in TC at t1, then desiring one option such that it becomes the effective intention, at t2, whereupon the other option ceases to be live. Anselm here is very clear that he is carving out new theoretical ground. He makes a point of coining a new term—*pervelle*—to capture the desiring "through" such that the desire is now an intention.[26] I will use "per-will"; a bit awkward perhaps, but it expresses the important technical concept that the choice for B is the *successful continuation of desiring B* to the end of the TC stage so it becomes an effective intention. At this point, given that pursuit of A and pursuit of B are mutually exclusive, the desire for A is overridden and ceases to be viable. As Anselm explains it, S (Satan) desires both to be just and to have the inappropriate benefit, then he "expels (*expulit*) his good desire by the evil surpassing it."[27] (Anselm doesn't say it, but perhaps a weak desire for A may linger on in a sort of wishful-thinking form. But, on the

flipping a coin and so randomizing whether or not one does that "best" that one decided on; *Brainstorms* (Ann Arbor, MI: Bradford Books, 1978): 295. This is, of course, nothing like the situation that Anselm envisions. Other philosophers have noted that libertarians would do well to locate the point of indeterminism *at* the choice rather than before it. See Balaguer (2010): 158–9.

[24] If a decision to postpone making a decision is a morally significant option, then the agent might be in TC regarding whether to decide now or later.

[25] DC 1.7. [26] DCD 3.

[27] DCD 3. See Volume I of Schmitt's edition of Anselm's *Opera Omnia*, p. 240, lines 7–9.

assumption that our *rational* agent understands that pursuit of B entails that he cannot pursue A, once committed to B, S must let go of his pursuit of A as a genuine option.) The motivating power in S's choosing B comes from the desire for B. But this is not an event-causal analysis. It is not the case that S's desire for B produces S's *opting* for B *over* A, even probabilistically. It is S himself who *expels* his (viable) desire for A by per-willing B.[28]

A Se Causation without "Special" Causal Powers

The *opting* for B *over* A comes from S. S really could have opted for A over B, by desiring A to the point where his desiring A "expels" his desiring B, and so the desire for A becomes the effective intention.[29] In that case the motive power in S's choosing A would have come from S willing and per-willing A. The metaphor of desire being "the wind in the sails" of an *a se* choice is helpful (though it is also misleading, as I will note) in distinguishing two causal "moments" in the choice. Take a sailboat with a fixed sail, so we change direction using a rudder. If the boat veers north there are (for our purposes) two main causal factors involved, the wind and the rudder. But it is only the wind that gives the boat the power to *move*. It is S's *desiring* B that causes S's choice for B in the sense of providing an active force, like the wind in the sails. The rudder does the different job of determining the direction. The *affectio* provides the motive power, but it is not what explains the opting for this over that.

On Anselm's analysis, the indeterminism in the *a se* choice comes in right at t2 when S opts for B over A. Thus Anselm's proposal is quite different from Kane's. On Kane's description, the indeterminism precedes the choice, and the choice itself is probabilistically caused as the termination of a non-determined process. Kane would have it that S, through a non-determined process, debates between A and B, then the motivation for B probabilistically causes him to opt for B. In that Kane associates this process with brain activity, the indeterminism in the debating process might be constituted by the non-determined behavior of particles in the brain, and the choice be a consequence of that behavior. It is easy to see why some philosophers hold that Kane renders a free choice just a matter of luck. It does seem "lucky" that an indeterminate neuronal process ends in the particles "bouncing" one way, when they could have "bounced" another.

[28] This description is echoed by psychologist Roy F. Baumeister (2008: 69). He writes that, "To the extent the free will exists, it serves not to initiate action so much as to alter and steer it." (Which is not to say that Baumeister is sympathetic to libertarianism per se.)

[29] This point should not be phrased as "the desire for A expels the desire for B" since that suggests that it is the desire, not the agent, which explains the preference.

Kane responds to this "luck" criticism by arguing that, although there is indeterminism in the process, if an agent intends an action, tries and succeeds in accomplishing it, and endorses it as his own, it is not a matter of luck. However, the examples he gives in making this case undermine his argument.[30] One example is the shooter who, due to wind conditions and whatnot, might or might not hit his target. Another is the husband who, hitting a table top, might or might not break it due to the properties of the table top. In both examples, the indeterminism involves something which happens after the choice and which is external to the conscious choosing of the agent. If the shooter succeeds in hitting his target and the husband succeeds in breaking the table top, the non-determined aspect of their achieving what they intend does seem a matter of luck in each case. The examples reinforce the suspicion that simply having some indeterminism in the process of choice does not contribute anything that could ground or enhance responsibility. Regarding the examples, Anselm would hold the agents responsible on the basis of what they intend, and the subsequent, probabilistically caused, effects would not be relevant to the moral status of the agent.[31] (The "probabilities" in Kane's examples of the shooter and the husband apparently refer to situations involving wind conditions and the tensile strength of the table top. I will argue, in discussing "luck" in Chapter Seven, that libertarians would do better to drop talk of probabilities altogether, if the probabilities in question refer to the likelihood of one option being chosen over the other.)

Unlike Kane, Anselm does not locate the relevant indeterminism in the process preceding the choice. The situation, as Anselm sees it in *De casu diaboli*, is that S's being in TC is in fact determined by God, since it was God who supplied the motivating *affectiones*, and this is S's first choice. S's effort (or efforts) to pursue the mutually exclusive options is causally necessitated by God. If God had provided the created agent with only one *affectio*, then his pursuing it would have been necessitated and unfree. By providing conflicting *affectiones*, God opens a small space for the created agent to choose on his own, and the indeterminism occurs at that moment of choice. Anselm grants that the created agent has only a little aseity. In Chapter One I noted that the libertarian might be accused of an unseemly pride when he claims that he is able to help in his own creation. But all the created agent does is to opt for this or that God-given *affectio*. There is nothing here to encourage a foolish pride. But there is just enough to allow the created agent to contribute, albeit in a small way, to his own creation.

[30] Robert Kane, "Responsibility, Luck and Chance." *Journal of Philosophy* 96 (1999): 217–40.

[31] How the shooter or the husband evaluate their success afterwards might itself be a moral choice and might have significant impact on their future characters.

Alfred Mele, criticizing Kane, argues that, since efforts motivating a choice may be imposed upon the agent, the indeterminism involved in the choice does not provide any control worth wanting.[32] Kane's analysis is susceptible to this criticism in that, for Kane, one of the preceding "efforts" is what produces the choice, albeit probabilistically. For Anselm, however, it is the agent who brings about the choice. (Mele's criticism does have *some* traction against Anselmian libertarianism. I take up the matter in Chapter Five.) It is entirely up to S to opt for B over A. S's *opting* for B over A cannot be caused or explained—even probabilistically—by S's preceding desire for B, since S also desired A and could equally well have opted for A over B.[33] Anselm insists that the members of S's contrast class, the angels who ended up choosing A, were, in all relevant respects, in the same situation as S. He cannot say otherwise, since he will not allow the difference in their choices to be the result of some preceding factor; a factor which could only have come from God.

Anselm sees the worry in insisting that there can be no explanation involving some preceding factor. "Why then did he [Satan] will? . . . Only because he willed. For this willing had no other cause by which it was by any means impelled or drawn, but it was its own efficient cause, and effect, if such a thing can be said."[34] When Anselm says that the willing is its own efficient cause and effect, one might suppose, were this all one had to go on, that Anselm was introducing an incoherent thought; that the choice per se, without the agent, is its own efficient cause. In that an efficient cause is what acts to bring something into being, this would seem to be proposing an impossible situation—a non-existent choice bringing itself into being. But in that Anselm is posing the question, "Why did Satan choose to sin?" I think it is clear that Anselm does not intend to abstract the agent from the choice. The point is that *S's* opting for B over A is not the product of previous, God-given, factors. This is an uncomfortable point, but better than pushing the responsibility for S's sin back to God, which would be the upshot if there were some preceding factor which explained the choice.

[32] Alfred Mele, *Free Will and Luck* (Oxford: Oxford University Press, 2006): 52.

[33] Kane can say the same, but then Kane's appeal to probabilistic causation as an attempt to give an explanation for a choice becomes problematic (1996: 176–9). A probabilistic cause makes an uncertain outcome more likely than if the cause had been absent. At t1 the desire for B does render the choice for B more likely than if it had not been present, but simultaneously the desire for A renders the choice for A more likely than if *it* had not been present. Without the desires neither choice could occur. But if the desire for B is a probabilistic cause which *explains* the choice for B *over* A, it is hard to see how the desire for A could be an equally viable, explanatory, probabilistic cause for the choice for A *over* B. More on this in Chapter Seven.

[34] DCD 27. (My translation; Schmitt I p. 275, ll.30–3.)

But if the actual opting for this over that is *not* produced by preceding events, but rather by the agent, can't Anselm be seen as a *standard* agent-causalist? In that Anselmian libertarianism insists that the preceding motivations provide the motive power, but also insists that the agent himself produces the choice, does it foreshadow Randolph Clarke's "integrated agent-causal" theory?[35] Anselm should be considered an agent-causalist, since he subscribes to an overall substance-causal view. Moreover, what he is concerned about from the beginning is where to locate some trace of aseity on the part of the rational created agent, such that that agent can be justly held responsible for his choices and have some role—however small—in contributing to his own character formation. The grounding for the agent's responsibility comes from the agent's opting for this over that, and it is absolutely up to the agent which *affectio* is per-willed and becomes the intention. So granting that Anselm is an agent-causalist properly emphasizes the locus of responsibility in his system.

But there is a key difference between Anselm's view and most contemporary agent-causal theories. Anselm's insistence that God is the source of everything with ontological status requires him to stop short of embracing what proves to be the central difficulty with most agent-causal theories. The defining point (and main target for criticism) of Clarke's view, and of other standard versions of agent-causation, is that they ascribe to the agent an additional and different causal power beyond the motivating factors preceding the choice. (This is part of that denaturalizing that Steward complains of.) Clarke writes, in discussing where event-causal theories are lacking, that what is needed are "further positive powers to causally influence which of the alternatives left open by previous events will be made actual."[36] Anselm allows that the agent does have "positive powers to causally influence which of the alternatives left open by previous events will be made actual." What he denies is that these are "*further*" positive powers, if—as in most agent-causal theories—"further" indicates some new sort of causation, that is not shared by lower animals and that is added to the powers associated with the motivating factors, the *affectiones* in "use." Anselm holds that S chooses B *a se*, but he does not introduce any new causal "force" worthy of the name.

Anselm argues that S opts for B over A, not through the exercise of a new power, but *by per-willing B.* At t1 S desires both A and B, and then, at t2, he desires B such that the desire for B becomes the intention, at which point the desire for A is overridden and ceases to be viable. As regards the continuum of the process of motivation, S's per-willing is similar to the horse's when it per-wills to eat, such that the desire to eat becomes an intention to go out and graze. The

[35] Clarke (2003): 133–49. [36] Clarke (2003): 108.

difference, of course, is that the horse must, by natural necessity, exercise its desire to graze. It could never be in TC, evaluating morally significant options. Even if we allow that it "chooses" between trivial options, it cannot be just or free.

On Anselm's analysis, God causes and sustains the *affectiones* for as long as they exist. When S is in TC, God is sustaining the desires for A and for B. When S per-wills B, forming the intention to B and "expelling" the desire to A, God sustains the intention to B. Had S per-willed A, God would have sustained the intention to A. God is the cause of everything with genuine ontological status in S's choice, but his choice is *a se* because it is absolutely up to S whether he per-wills A or B.[37] Satan, for example, "threw away" his original desire for justice, not because God failed to give him help without which the fall would be inevitable, but because, on his own, he chose to pursue the inappropriate benefit, when he could have chosen to maintain justice.

But isn't the choice for B over A a new event in the world? Isn't it something in addition to S's God-given *affectio* for B? (Remember that neither B nor the desire for B are intrinsically evil. Rather B is something that S ought not to will at the time, and the evil comes from the disobedience which entails the rejection of justice.)[38] So don't we have to attribute to S some new causal power? The metaphor of the sailboat veering north appealed to two distinct, and both very real, causes, the wind in the sails and the rudder. If the motivating *affectiones* supply the motive power, the "wind in the sails," for a choice, mustn't there be some additional cause analogous to the rudder or, at least, to its turning?

The Anselmian answers that the sailboat metaphor fails at this point. The choice for B *is* an event; a happening at a time. But it is not an event that introduces anything new, with ontological status, to the sum of what exists in the universe, and so it need not be the effect of some new causal power. It is just the point in the process of choosing when S per-wills B, so that A ceases to be a live option. And so it does not require any more cause than S's per-willing B.[39] In a creaturely choice, God is the cause of all that *exists*, but He is not the cause of all that *happens*. (The claim that a choice is not any sort of a "thing" may seem puzzling prima facie, depending on your views on ontology. I develop this

[37] Anselm allows that created agents have an impact on what God does, a position that was anathema to many of the great medieval philosophers, including Augustine and Aquinas. See Rogers (2008): 120–1.

[38] DCD 4.

[39] In Rogers (2008: 59–60) I had said that, on the Anselmian analysis, created free agency should be understood to be "primary" rather than "secondary" causation. But primary causation is God's act of causing things to *exist*, so "primary" causation is the wrong term for human agency.

entailment of Anselmian libertarianism a bit more in Chapter Four and discuss some implications.)

Thus Anselm solves the puzzle of how choices can be up to created agents, without proposing any new causal powers in the universe. As an agent-causal view, the Anselmian approach holds that the choice is absolutely up to the agent, and so it avoids some of the problems generated by Kane's event-causation. Kane described the choice as the terminus of a non-determined process, where preceding "efforts" probabilistically cause the choice. This allows for indeterminism, but it is hard to see how the agent himself is the source of the choice. On the other hand, as opposed to most standard agent-causal theories, Anselmian libertarianism does not introduce any new causes to explain the agent's choice. The agent chooses just by per-willing, period. So, while Anselmian libertarianism is a version of agent-causation, in its caution not to multiply causes it is quite different from most other agent-causal theories.

One who already embraces a robust agent-causal theory may hold that something is lacking; that without the exercise of a special agent power producing a new *thing*, the choice, the Anselmian agent's control is too "slender" to bear the weight of moral responsibility. Moreover, the agent-causalist may continue, we need the additional "oomph" supplied by the special agent-causal power to explain how and why the choice occurs. Were Anselm confronted with this claim, he would have to respond that, if there were such a "special" power, it must come from God, and if its exercise constituted some new *thing* in the world, then that, too, must come from God. Thus, for Anselm, insisting that the choice is an existent "something," and is the effect of a robust agent-causal power, assigns all the determinative power in a choice to God and undermines the effort to find some small space for created aseity.

Suppose we depart from Anselm's theism and propose that it is not God *and* the natural universe, but only the natural universe that brings things into being.[40] The agent-causalist must then say that the agent-causal power is a product of the natural universe alone. He can then suppose—since we have set aside theological commitments—that the choice is some real *thing* with ontological status and is simply the non-determined effect of the agent's exercising his power. Prima facie this thesis may seem to provide a better grounding for moral responsibility, in that what the agent bears responsibility for, the event of the choice, has a more

[40] There are a plethora of other alternatives—a deist god designs the "natural" universe to produce agent-causation; our universe is a video game and the game-players design it to include agent-causation, etc. etc.—but most of the contemporary free will discussion proceeds without the assumption of a designer for the universe, so the most obvious alternative to Anselm's theism is a non-theistic naturalism.

robust sort of existence than on the Anselmian theory. But now the opponent of agent-causation may fairly say that the view seems *more* mysterious than ever in that it proposes that the human agent, through a naturally produced power, brings something new, the choice, into being *ex nihilo*.

The standard agent-causalist may respond that the opponent here is just adding a new flourish to the same old criticism regarding the mysterious nature of agent-causation. The opponent, though, can say that it is indeed a *new* point, in that the ontological status of the act of choice has not been discussed in the literature before. The standard agent-causalist may reply that there is nothing uniquely worrisome about the free agent bringing the choice into being *ex nihilo*. Doesn't any creative human activity involve bringing new things into being? But the best answer to that rhetorical question is probably just "No." Our other causal activities, no matter how creative, consist in rearranging the elements that are given to us in the universe. The most inventive video game, the most surprising literary character, are constructed out of previously existing elements. We can arrange and rearrange what is given to us in the universe, but we do not actually produce any new thing. On Anselm's analysis our choice is a new event, but it is "made of" preexisting elements.

The standard agent-causalist may respond that the Anselmian begs the question here; he may argue that the point of the agent-causal theory is that we do indeed produce a new, actually existent something with ontological status—the choice—out of nothing. That is, we add to the overall sum of what there is in the universe. But in that case, this version of agent-causation seems doubly mysterious. Not only has the natural universe thrown up a new and special power by which free human agents choose, but this power allows free human agents to do something which, if not quite impossible, has been traditionally reserved to God—we can now bring new things into being *ex nihilo*. In the past, some leveled similar-sounding criticisms at libertarianism in general. If the adherent of standard agent-causation chooses to insist that the event of the choice must be understood as a new thing with ontological status, then this old criticism does seem to stick against his theory.

Nor does the standard agent-causal view add anything which renders the choice more fully explicable than on the Anselmian view. If our question is, "How and why did S choose B rather than A?" the Anselmian responds that the "how" is answered, "by per-willing B," and the "why" is answered, "just because B is what he per-willed, rather than A." (Of course he can give reasons for the choice for B, but he could have given equally explanatory reasons for the choice for A.) That is the best the Anselmian can do, since he cannot appeal to preceding properties or events as the cause or explanation of S's opting for B over

A. On the standard agent-causal view the "how" is answered, "by agent-causing B over A." This is a different answer than the Anselmian one, but it does not seem to be any more complete or satisfying an explanation. "Per-willing" is just replaced with "agent-causing." And the "why" is answered, "just because B is what he agent-caused, rather than A." The standard agent-causalist has added a new causal power, where Anselm has not, but the nature of the causal power is suspicious and its addition does not enhance our understanding of the agent's choice. If libertarians who are concerned about aseity can get by with Anselm's more parsimonious approach, better to do so.

Phenomenologically, Anselm's description in which the agent struggles between options and ultimately pursues one or the other seems adequate. It is surely not obvious that we *experience* the exercise of some further control when we make a decision. And, theoretically, the limited *aseity* Anselm proposes, which can be secured without the introduction of new things and principles to the universe, may be sufficient to ground responsibility.[41] In a real human life, if one builds one's character slowly over time through many *a se* choices, perhaps one's autonomy and control would grow over time. This is one avenue to explore in trying to offer a more developed Anselmian libertarianism.

[41] If the *aseity* Anselm defends seems very limited, note that in Anselm's tradition of classical theism the proposal that we, and not God, cause some events, is a radical thing to say.

4

Three Entailments

Introduction

In this chapter I will discuss three entailments of Anselmian libertarianism. In developing the theory, none of these entailments requires a chapter of its own, but nonetheless it is important to have them in mind early on, as they will come up over and over. The first entailment is the point noted towards the end of the last chapter, that a consequence of Anselm's theory is that the act of choice is not a discreet "thing," using "thing" in a somewhat specialized way to label a robust existent with ontological status. A choice is not something over and above the motivations that provide the causal power for the choice. I argue that, although prima facie this might seem an odd claim, it is actually plausible theoretically and phenomenologically. I suggest that it is good for philosophers engaged in the free will debate to keep in mind the possibility that the choice is not a thing. This is especially true insofar as the work of experimental psychologists impinges on the contemporary discussion.

The second entailment is that what grounds the truth of a proposition concerning an *a se* choice, and hence what is necessary for any knowledge of the choice, is the actual making of the choice itself. I call this the "grounding principle." Some who call themselves libertarians deny this principle. And some suggest it without making it as clear and unequivocal as it is in Anselm's theory. For Anselm, as explained in Chapter Three, the aseity of the created agent is rooted in being able to choose between alternative possibilities. The need for open options has been challenged in recent decades through Frankfurt-style counterexamples. In Chapter Six I argue that such counterexamples cannot even be framed within Anselmian libertarianism, and the grounding principle plays a significant role in that argument. A discussion of the grounding principle also helps to develop the important concept of external non-causal (ENC) necessity and show why it conflicts with aseity.

The third entailment of Anselm's theory is that *a se* choices are our building blocks as we construct our own characters. The thought that our choices and actions help to develop our characters is at least as old as Aristotle, and is embraced by many participants in the free will debate. It is helpful to offer a brief outline of this aspect of Anselmian libertarianism early in the book, since it will come up periodically. It is especially important in Chapter Eight where I claim that the Anselmian has a way to mitigate the luck problem, based on emphasizing the character-forming nature of choice. This third entailment is also relevant to the "tracing" issue. Anselm holds—and many contemporary philosophers say something similar—that since you are responsible for your *a se* choices, you are responsible for the character they create. This means that, once you have begun to construct your character, you can be responsible for a choice determined by your character. Recently an interesting problem has been raised against this tracing thesis: (To cast it in Anselmian terms) As you are making the *a se* choices which form your character you do not know what choices will confront you in the future, and perhaps you do not know that you are forming your character by your choices. But if you don't know what the consequences of your choice will be, can you truly be responsible for the character that is formed and the choices determined by that character? If the answer to this question were "No!" then Anselm's (and the Anselmian's) entire project is undermined. In Chapter Nine I try to answer this problem with the tracing thesis. The first entailment mentioned above, Anselm's point about the ontological status of the choice, will play an interesting role in developing this answer.

The Ontological Status of Choice

> No creature has anything (*aliquid*) from itself (*a se*). It doesn't even have itself from itself, so how could it have anything from itself? For, if there is only one [being, i.e. God] who creates (*fecit*), and whatever is made is from that one, it is clear that by no means can there be anything except the one who makes, and what He makes.
>
> *De casu diaboli* 1

That a choice—as distinct from the *affectiones* that motivate it—is not a discreet being with ontological status is exactly the conclusion that Anselm himself hopes to defend.[1] He is confronted with what can be called the dilemma of created freedom and divine omnipotence. All that has real existence comes from God, yet human agents can make choices from themselves. As I noted in the previous

[1] This section draws on my "Anselm on the Ontological Status of Choice." *International Philosophical Quarterly* 52 (2012a): 183–97. There I make the textual case that Anselm does indeed consider a choice, per se, not to have ontological status.

chapter, Robert Kane attempts to solve an analogous problem: How can we explain freedom, in the sense of an agent's having ultimate responsibility for his choices, without introducing any new powers or entities outside the sciences and specially invoked to allow for that freedom? Kane himself, though, does not address the question of the ontological status of choice. Nor do the agent-causalists. In its most general formulation the dilemma which both Kane and Anselm hope to solve is how to defend freedom (i.e. Kane's ultimate responsibility or Anselm's aseity) without positing new entities, entities not already "given" by science or God (or both). Call this the dilemma of freedom and parsimony.

Anselm solves the dilemma by theorizing that the rational created agent possesses the two *affectiones* that can come into conflict when someone desires justice, but also desires some benefit that falls outside of the set of appropriate benefits. God is the creator of the agent, his will as instrument, his *affectiones*, and so God is the source of even the use of the will as an exercise of the faculty through the motive power provided by the *affectio*. Hence God is the creator of all that has ontological status in the use of the will, since what that use *is* is just the following of the *affectio* in a particular instance. Even in sin, the will as instrument and the *affectio* for benefit—in this case directed towards an inappropriate benefit—are from God. In a free choice, all that is up to the created agent is that it follows its inclination for justice rather than for inappropriate benefit, or its inclination for inappropriate benefit rather than for justice. And the "following-this-rather-than-that..." does not constitute any *new thing* added to the sum of what there is in the universe. So God causes all that *exists*, but He does not cause all that *happens*. He does not cause the free choices of created rational agents.

The claim is that whatever in a choice has ontological status comes from God. Nonetheless, that the choice *happens* as it does is up to the created agent such that the *preferring* of one option over another does not have any independent ontological status. In much of the free will literature, classical and contemporary, a choice is treated as separate from whatever motivating factors are relevant to its history. It is some discreet and unique act that can be isolated from what precedes and follows it, and considered in itself. Here I want to argue that Anselm and the Anselmian see it differently and that their position has plausibility. The claim is that there just is no such *thing* as a choice understood as an action separate from the motivating inclination and having its own ontological status.

In italicizing "thing" I indicate that my use of the term may be somewhat specialized and, I fear, a bit intuitive. The impetus behind the Anselmian usage is that—as the quote at the beginning of section indicates—any "thing," *aliquid*, is made to exist by God. It might be argued that if some x can be labelled, given a name, it ought to be considered a thing. Perhaps that is a standard contemporary approach, but it is clearly not Anselm's view. We can speak, for example, of "injustice," indeed we can point to specific instances of injustice. But Anselm has no doubt that injustice is not from God. As simply a lack of the justice that ought to be there, injustice is not a thing on the Anselmian understanding.[2] Similarly, Anselm's use of "thing" is at odds with the thought that if some x can serve as the subject of a sentence, or the value of a variable, then x must be a thing. "Injustice is rampant in the world today," seems a well-formed sentence, and yet Anselm would hold that injustice is an absence, not an *aliquid*. By the same reasoning, not every property is a thing. On Anselm's account, good properties are things and come from God. But, for example, the negative property of "lacking justice" does not come from God and is not a thing. Nor is every relationship a thing, by this Anselmian usage. Some are, such as the relationships which constitute the three Persons of the Holy Trinity.[3] But some are not. Say that this murder was more unjust than that robbery. The relationship of being "more unjust" does not constitute some existent in the universe.

True, one might argue that the above analysis does not apply to a choice, per se, since a choice is obviously not a lack or an absence. It is a genuine event. And Anselm is insistent that all that *exists* in the event is from God. But, as I argued in Chapter Three, he is equally clear that God does not cause the choice itself. He does not cause the choice to sin, nor does He prevent it. Preventing sin in the context of created agency would be causing the choice to hold fast to the good, thereby destroying the moral value of the "choice." So God does not cause the choice either way. In that God causes whatever can be said to exist, it follows that my locution—choice per se is not a thing—is appropriate in the context. (If the term "being" or "entity" or "existent" better captures my suggestion for what an Anselmian choice is *not*, then "thing" can be replaced by one of those terms.)

Granted, prima facie it seems odd to claim that a choice is not a thing. Is there more to say to develop and defend the view? Unfortunately Anselm himself does not elaborate on the ontological status of choices, or actions, or events, beyond holding that God causes what exists in a free choice, but that He does not cause the choice itself.[4] A glance at contemporary work on the ontology of actions or events is not especially helpful for our purposes here.[5] Recent overviews of the

[2] *De casu diaboli* (DCD) 9. [3] *De Incarnatione Verbi.*

[4] An obvious source to check, in that we know it influenced Anselm's work, is Boethius' commentary on Aristotle's *Categories*. Alas, the section on actions is woefully brief, that chapter of the *Categories* having been left unfinished. And that means, of course, that we cannot even turn to Aristotle for help.

[5] But for a brief overview see E. J. Lowe, "Action Theory and Ontology." In *A Companion to the Philosophy of Action* edited by Timothy O'Connor and Constantine Sandis (Chichester, UK:

issue often begin with the observation that the subject is young.[6] But I think it is appropriate to understand a choice as what might be called a "thin" event. It is a happening at a time, but it is ontologically dependent on the elements producing the event in a way that allows the event itself, as distinguished from those elements, not to have ontological status.[7]

An analogy may help. (It is an analogy inspired by one Anselm proposes in talking about *voluntas* (will) as an instrument, an ability, and a use.[8]) Note that the elements in the analogy are not intended to correspond to the entities in an *a se* choice. All I hope for here is to defend the thought of a "thin" event. Suppose that M challenges R to a foot race. R is less athletic than M, and so M, being sporting, gives R a head start. Off they go. Their bodies are certainly things with ontological status. It does not seem too much of a stretch to consider their strength or power for running to be things. If we are very liberal in our ascription of ontological status, even the running itself, as the exercise of the power belonging to the instrument, may be a thing. But now suppose that M overtakes and passes R. The passing is an event. We can refer to it; we might say, for example, "M's passing R allowed him to win the race." We can assign a time to it; "M's passing R happened at 3:15:22 precisely." It may even be an event with important consequences; "M's passing R, which allowed him to win the race, meant he won the million dollars." But the "passing" seems even less thing-like than the running. Yes it is a relationship, and if we are committed to being extravagantly liberal in our ascriptions of ontological status, perhaps we may call it a thing. But it seems entirely appropriate to say that, while the exercise of the power of running may be a sort of thing—perhaps as an extension of the strength or ability in question—the passing just is not a thing. It is an event, but it is a thin event, having no ontological status distinct from the "runnings" of M and R.

I suggest that an *a se* choice such as our standard example, S's choice of B over A, can be described as a thin event. It is S per-willing B, where S, his faculty of

Wiley-Blackwell, 2010): 3–9. Lowe notes that according to Quine's dictum that "to be is to be the value of a variable," events and actions ought to be said to have ontological status (3–4).

[6] e.g. Peter Simon, "Events." In *The Oxford Handbook of Metaphysics* (Oxford: Oxford University Press, 2003): 357–85. Perhaps the later medievals had some insightful things to say on the topic, but a study of their work is not necessary for the purpose of developing Anselmian libertarianism.

[7] I would describe the "thin" event as "supervening" on these elements, were it not that Brian McLaughlin and Karen Bennett say quite firmly, in the "Supervenience" entry in the Stanford Encyclopedia of Philosophy that, "Supervenience is not a relation of ontological priority," *Supervenience* 3.5. What I am going for with the suggestion of a "thin" event is precisely a relation of ontological dependence.

[8] *De Concordia* (DC) 3.11.

will, and his desire for B, all come from God (or nature or both). The per-willing which constitutes the choice occurs at the point where S's desiring B makes it the case that his desiring A is no longer viable. S "overrides" the desire for A by per-willing B. But the overriding is no new thing added to the universe. S's choice for B is *a se* because he really could have chosen A, and it is from himself that he chooses B rather than A. One could suggest that the choice for B possesses the odd property of "rather-than-A-ness," but surely "rather-than-ness" is not a "something" either.[9] S has not brought some *potiusitas* into being. (*Potius* being the Latin for "rather than." Metaphysics seems to cry out for the invention of barbarous Latin terms.) So S's *a se* choice of B just does not bring anything with ontological status into being.

In the literature, libertarian choosing is sometimes portrayed as the achievement of a sort of stasis, a balance between competing desires, followed by a new and sudden event, coming out of the blue, The Choice. But this is not the picture Anselm presents. There is the struggle to exercise the two *affectiones* and pursue conflicting desires, and there is the ultimate success in pursuing one over the other. Which desire is followed is up to the created agent alone. It is the agent who wills *this* to completion and, in so doing, fails to will *that* to completion. And the point at which the agent per-wills *this* is the choice, a thin event.

Some objections to this portrayal of a choice come to mind. The Anselmian claim is that the choice just *is* the desire or inclination pursued to a certain point, the point where it becomes an intention. But then the critic might note the obvious fact that one can desire something without actually *choosing* it. Doesn't this show that there must be a distinction between desire and choice, and would not such a distinction undermine the Anselmian thesis? I noted in Chapter Two that Anselm—pioneer that he is—does not make many of the distinctions which the contemporary philosopher finds important. He does not distinguish as clearly as one might wish between a desire and an intention, referring to both as *affectiones*. Nor does he distinguish between distal and proximal intentions. But his system does include the key distinction between what we might call "mere" desire, and desire per-willed to the point of intention, where it becomes choice. The claim that the choice just *is* the desire carried to a certain point does not conflict with this distinction. To offer a supporting analogy: You can, for example, drink without drinking to excess, but drinking to excess is nothing over and above drinking.

[9] Someone who has a taste for flouting parsimony might like the thought that every property—this red ball's not being blue, it's not being a toad, it's not being omniscient, etc. etc. ad infinitum—is a thing, but it is hard to see how that helps us get on with the business of metaphysics.

But must not the choice be a new action? A sort of "setting oneself to act"? A concrete volition which is separate from the motivating factors which led up to it? Unlike the motivating factors, which are in play throughout a process (and may even exist in an agent non-occurrently in Anselm's schema) the choice is something that happens at a time. Doesn't this mark it off as an additional "something" in the world? Not if my suggestion above, that choice is a thin event, is plausible. Moreover, there is good reason not to allow the description of choice as a "setting oneself to act." The choice itself is plausibly labeled an act. It is an event brought about by an agent. On the Anselmian account, insofar as it is the pursuing of one desire over another, it has the ontological status of the desire. But if there is a separate "setting" of oneself to act, then that, too must be an act. Presumably, if the agent is to be free and responsible, that act of "setting" must be the product of a free choice, a "setting oneself to act." And we open the door to an infinite regress.[10] We avoid the problem if we eschew the temptation to think of the choice as a new action, ontologically separate from the motivating factors that came before.

Perhaps the critic will say that the Anselmian approach conflicts with experience. Don't we experience making a choice as something different from the motivating factors that came before? Even if we do, we might be experiencing the success at pursuing one inclination over another, just as we might experience the passing in the race without that entailing that the passing is some new thing. And it is not so clear that we actually do experience that success considered as a recognizable event at a discreet time. Surprisingly little work has been done on the phenomenology of choice.[11] I find that, when I try to think back to the actual experience of making some morally significant decision, the whole process is much more chaotic and fuzzy than the philosophical literature would make it out to be. This is not a criticism of the literature. In order to get at the essentials, one has to simplify. This is true of the present work, and it is certainly true of Anselm's own explication in which my development of the Anselmian theory is rooted. My point is just that, to date, the literature has dealt more with the theoretical issues than with the actual experience of choice, and it may be that the impression that there must be a distinct act of choice, separate from the motivating factors, is more a product of the theoretical discussion than of our genuine memories of what we

[10] E. J. Lowe, *Personal Agency* (Oxford: Oxford University Press, 2008): 7–8.

[11] The Libet experiments and the literature that has grown up around them suggest that studying choices—both from the perspective of the first person and that of the third person—is a difficult task. For an overview of the Libet experiments see Benjamin Libet, "Do We Have Free Will." In *The Oxford Handbook of Metaphysics* (First Edition) edited by Robert Kane (Oxford: Oxford University Press, 2002): 551–64.

experienced when we made a choice. In my own chaotic and fuzzy memories, the struggle to decide what to do looms large, while a "moment" of choice is not discernible. I do not rule out the possibility that my memory is weaker than most, but if my memory reflects a fairly standard experience, then it is not clear that the phenomenology of free will undermines Anselm's thesis.

I believe that consideration of this entailment of the Anselmian theory—that a robustly free, *a se* choice is a thin event—has an important place in the contemporary debate about free will. Many philosophers are looking now at the evidence which certain experimental psychologists adduce to claim that they have demonstrated empirically that human beings do not have free will. Of course, the psychologists haven't demonstrated that at all. But what is shocking from the philosopher's perspective is that these experimental psychologists do not explain what they mean by "free will." Some of the free will experiments in question—ignoring the need for conceptual spadework—depend upon assuming that there is a readily observable temporal location for The Choice. Then they (supposedly) pinpoint such a location, and then observe what, in consciousness and/or brain activity, leads up to it, what is simultaneous with it, and what comes after it.[12]

This same point, that it is crucial to consider just what sort of an event a choice is, is also relevant to whatever neurobiological work is done on freedom, choice, and will in the future. Mark Balaguer, who is optimistic about the ability of science to establish whether choices can be undetermined, writes, "the question of whether libertarianism is true just reduces to the question of whether some of our torn decisions are undetermined in the appropriate way. This, of course, is a straightforward empirical question about the neural events that are our torn decisions."[13] But in attempting to observe these neural events, it must surely be helpful to think about what sort of events they are. Suppose that when it comes to the sort of choice in which we are really interested—the morally significant choice which has always driven the free will discussion—Anselm is right; as something separate from the motivating factors, there is no discreet *thingish* sort of event to be experienced, isolated, and studied. The Anselmian account would not rule out the possibility of establishing that there is a "thin" event that occurs at a "moment of choice," but if we find Anselmianism plausible, that should affect how we think about just what it is we are looking for and how we go about the search. At the very least, it is well to note that what actually constitutes a choice deserves more discussion than it has received in the literature on studying free will within the sciences.

[12] The Libet experiments take this form. Benjamin Libet himself does not conclude that we are not free, but his experiments are often cited as evidence for that conclusion.
[13] Balaguer (2010): 69.

Anselm's thesis about the ontological status of choice provides a workable solution to the puzzle he himself is trying to solve, the dilemma of created freedom and divine omnipotence, and also to the problem stated more generally as the dilemma of freedom and parsimony. If Anselmian libertarianism can give us aseity without multiplying entities, that is a distinct point in its favor. And, in addition to this theoretical advantage, it also suggests a hypothesis that needs to be considered in the contemporary discussion involving the experimental approach to the question of human freedom.

The Grounding Principle

> [Granted, divine foreknowledge does entail a kind of necessity]...but this necessity follows upon (*sequitur*) the positing of the thing, it does not precede it. It means the same as if we were to say, "What will be, by necessity will be". This necessity signifies nothing else than that if something will be, it cannot at the same time not be.
>
> *De Concordia* 1.2

A second entailment of Anselmian libertarianism is this: in that an *a se* choice comes from the agent himself in the most absolute way, the truth of a proposition about that choice must be grounded in the making of the actual choice. I use the term "grounded in" in a broad sense that does not express any very fine-grained understanding of just what the relationship between a proposition and the state of affairs it purports to describe is. A discussion of that issue would take us too far afield, and is not necessary for the elaboration of Anselmian libertarianism. Those who prefer to talk about "truth-makers" can say that it is the actual choice that is the truth-maker for the proposition about that choice, and others are free to adopt their favored locutions. The point is that the Anselmian commitment to aseity entails that the truth value of the proposition about the free choice depends on the making of the actual choice. There is nothing but the actual choice which can be the ground for the truth of the proposition about it, and there is nothing but the actual choice which can be the origin—causal or otherwise—of knowledge about the choice. Taking our example in which S chooses B at t2, the truth of "S chooses B at t2" is grounded in S's *actually choosing* B at t2, and any knowledge that S chooses B at t2 originates with S's *actually choosing* B at t2.[14]

[14] Thomas Nagle worries that agents finds themselves in situations over which they have little control. And then, "We judge people for what they actually do or fail to do, not just for what they would have done if circumstances had been different." He finds this an aspect of responsibility-undermining moral luck. The Anselmian responds that there is no "what they would have done." Yes, people find themselves in different situations, and are held responsible based on what they

Some contemporary libertarian philosophers seem to accept or assume something like the grounding principle.[15] But they do not spell it out and embrace it as clearly as Anselm does.[16] Anselm has to confront the issue because he is worried about the dilemma of freedom and divine foreknowledge, which was, and is still, a standard problem for those who want to posit free will in the universe of classical theism. A look at his solution, and an attempted solution which the Anselmian rejects, will make the principle clear and demonstrate its importance, and will also help to develop the Anselmian conception of aseity. The grounding principle will figure prominently in responding to Frankfurt-style counterexamples in Chapter Six.

The dilemma of freedom and divine foreknowledge is this: If God knows at t1, that S chooses B at t2, then, at t2, S cannot *actually* choose other than B at t2. He may be able to choose otherwise in some compatibilist sense—if he had wanted A more, then he could have chosen A, etc. etc.—but he couldn't actually choose otherwise. And, prima facie, if he couldn't actually choose otherwise, doesn't this show that God's foreknowledge conflicts with S's free choice? And so for all future choices. If God has foreknowledge of all choices, they cannot be robustly free.

Centuries before Anselm ever came to the issue, Augustine had made the helpful point that mere knowledge is not causal necessitation. A little later Boethius advanced the discussion by noting that we need to consider how it is that God knows the future. One possibility is that there is universal, natural determinism. All events at a time are causally necessitated by the state of the universe preceding that time. So God, knowing all the facts and natural laws in the present, deduces all that happens in the future. But then, says Boethius, the future choices are naturally necessitated, and hence they are not free. Instead, says Boethius, we should recognize that God is eternal and sees all of time as if it were present to Him. But Boethius is not a libertarian. He makes it clear that God

actually choose. It is difficult to see what Nagle might have in mind as a better universe for grounding responsibility; a universe in which all of us are in the same situation? (Thomas Nagle, "Moral Luck." In *Mortal Questions* (New York: Cambridge University Press, 1979): 24–38).

[15] See proponents of the "dilemma defense" defending the Principle of Alternative Possibilities against Frankfurt-style counterexamples; Carl Ginet, "In Defense of the Principle of Alternative Possibilities: Why I Don't Find Frankfurt's Argument Convincing." *Philosophical Perspectives* 10 (1996): 403–17; Robert Kane, *The Significance of Free Will* (Oxford: Oxford University Press, 1996): 142–4; David Widerker, "Libertarianism and Frankfurt's Attack on the Principle of Alternative Possibilities." *Philosophical Review* 104 (1995): 247–61.

[16] An exception is Kevin Timpe who does explicitly endorse a version of this principle, "Free Will: Alternatives and Sources." In *Philosophy Through Science Fiction* edited by Ryan Nichols, Fred Miller, and Nicholas D. Smith (New York: Routledge, 2008a): 397–408.

knows future choices as if they were present to Him because He knows that He is going to cause them. He is what I am calling a theist compatibilist. He denies natural causal determinism, since he believes that that would conflict with our free will. But he holds that God is the cause of our choices, and that that does not conflict with our free will.[17]

Anselm cannot make this Boethian move of arguing that God knows future free choices through knowing Himself and what He intends to do. On Anselm's analysis, while God sets up the situation at t1 such that S faces the open option to choose A or to choose B, that S chooses B *over* A at t2 does not depend on God in any way at all. God's knowledge at t1 that "S chooses B at t2" does not arise from knowledge that He has about Himself, nor about the state of the universe preceding t2. If S's choice is really *a se* then it is only the actual making of the choice itself that can ground knowledge of the choice. It is up to S at t2 to make it the case that God, at t1, knows that S chooses B at t2. And that means that God at t1 must have some sort of access to the *actual* choice for B at t2. How is that possible?

Anselm makes the simple and elegant move of introducing a new theory of time and eternity.[18] Boethius had said that God sees the future *as if* it were present to Him. Anselm replaces the "as if" with "because it *is* present to Him." Anselm adopts what can be termed the isotemporalist theory of time.[19] The view is that all times—what we limited human beings perceive, and refer to, as past, present, and future—actually exist "equally." (I am told that isotemporalism is common among contemporary physicists.) What is past, present, or future is subjective, relative to the temporal perceiver at a given time. In fact, all times have equal ontological status, and it is just our limited perspective that makes it seem as if all there is is the present. God, of course, sees things as they are, and so He sees all of time as equally "now" for Him, just as He sees all space as equally "here" for Him.[20] God can know, at t1, that S chooses B at t2, because He sees t1,

[17] Rogers (2008): 152–6. Hugh McCann (1995) proposes a position similar to the one I here attribute to Boethius, and holds, correctly I believe, that this was Thomas Aquinas' view as well.

[18] Augustine and Boethius sometimes say things suggestive of this view, but they do not spell it out clearly, and they do not use it to solve the dilemma of freedom and foreknowledge. They don't need to, since neither is a libertarian. See Rogers (2008): 158–68.

[19] This seems a better term than "Four-dimensionalism," in that one does not want to be dogmatic about how many dimensions there may be. And it is a better term than "eternalism" in that the term "eternal" has been used for at least the last two millennia to describe *God's* mode of being. The Anselmian claim is not that, with respect to time, the universe exists in the same way that God exists. I thank Catherine and Michael Tkacz for the term.

[20] The divine "seeing" is not like our perception. Nor does God know everything through knowing propositions. God's knowledge and His power are the same, and God "sees" by keeping all that has ontological status in being from moment to moment by thinking it.

t2, and all times, in one, eternal act. (This is not to suggest that God exists *at t1* in the sense of being limited to experiencing time as a sequence the way we do. It is to say that, if God exists, a temporal creature might truly say, at t1, "God exists.") So it is S's choosing B at t2 that makes it the case that God knows at t1 that S chooses B at t2. Anselm grants that, given divine foreknowledge, there is a sort of necessity to be ascribed to the foreknown choice. If God foreknows at t1 that S chooses B at t2, then necessarily S chooses B at t2. But this is a consequent necessity which follows from S choosing B at t2. It is exactly the same sort of necessity that applies to a present choice. If now you are reading, necessarily, now you are reading. You cannot now possibly fail to be reading. But Anselm argues that this consequent necessity is entirely innocuous since it originates with the choice of the agent himself. It is S's choice for B at t2 that is the source for the prior consequent necessity of "S chooses B at t2." So this consequent necessity does not undermine, or in any way conflict with, the aseity of the choice. Adopting isotemporalism allows Anselm to embrace divine foreknowledge of free choices, without loosening his grip on what he takes to be the core of free will, the aseity of the agent.[21]

But is this solution to the freedom and foreknowledge dilemma coherent? The Anselmian claim is that, given isotemporalism, S's choice of B at t2 makes the proposition "S chooses B at t2" true at all times. One might say that S's choice of B at t2 is "ontologically necessary"—it is a determinate part of reality—at all times. (But note, as will become clear, "ontological" necessity must be distinguished from ENC necessity.) But then, that S *not* choose B at t2 is ontologically impossible at t1. And that sounds ominous for S's ability to choose between open options at t1. If S's choice for B at t2 is *already* a part of determinate reality at t1, how is it that S confronts alternative possibilities such that it is truly up to S that he chooses B? The Anselmian grants that S's choice of B at t2 is ontologically necessary at all times. But it is S's choice for B at t2 that makes S's choice for B at t2 a part of determinate reality at all times. There is absolutely nothing independent of S's own actual choosing that makes it the case that S chooses B at t2. This is what aseity entails, according to the Anselmian.[22]

To underscore this point, compare Anselm's solution to the freedom and foreknowledge dilemma to a different solution which is popular among

[21] Rogers (2008): 173–84.

[22] Jeffrey Green and Katherin Rogers, "Time, Foreknowledge, and Alternative Possibilities." *Religious Studies* 48 (2012): 151–64. Trenton Merricks offers a related argument against fatalism: "Truth and freedom." *Philosophical Review* 118 (2009): 29–57.

philosophers of religion nowadays, Molinism.[23] The Molinist argues that God has foreknowledge of future free choices—free in a libertarian sense—because He possesses a sort of knowledge "in between" knowledge of His own nature and knowledge of what He intends to do. This "in between" or "middle" knowledge is knowledge of what are nowadays called "counterfactuals of freedom." These are propositions about what any possible, libertarian, free agent, would freely choose in any possible situation. Whether or not S ever actually exists, there exists a true proposition, "In situation Q, S would choose B at t2." God can foreknow that, in situation Q, S will choose B at t2 because God knows this true proposition, and if He adds knowledge of what He Himself will do, that is, He will create S and place S in situation Q, then He knows that S will choose B at t2.[24]

Supposedly Molinism has the advantage of reconciling not only freedom and divine foreknowledge but also freedom and divine sovereignty. Knowing all the counterfactuals of freedom, God can choose to make the mutually consistent world which suits Him best. The Anselmian is unconvinced, as any adherent of classical theism ought to be. Classical theism holds that what there is is God and what God creates. As we saw in the previous section on the ontological status of choice, positing something which exists independently of God restricts His power, hence the dilemma of freedom and divine omnipotence. On classical theism, even necessary truths like the laws of logic and mathematics do not exist independently of God. They are a reflection or entailment of His nature. In knowing Himself, God knows these truths. On the Molinist view God must take into account these independently existing counterfactuals as He goes about His business. They limit what He can accomplish. (The Anselmian allows that God must take into account the free choices of actual agents, *but He created the agents.* They do not just exist independently of Him.)[25]

Setting aside the theological question, positing these counterfactuals of freedom undermines Anselmian libertarianism in two ways (or perhaps it is better to say that there are two ways of putting the conflict between Molinism and Anselmian libertarianism). Most obviously, if it is absolutely up to S whether or not he chooses A or B, then it is not possible that there should be a truth to whether or not S *would* choose B, in the absence of an actual S and his actual

[23] For an extended look at the present Molinist debate, see *Molinism: The Contemporary Debate* edited by Ken Perszyk (Oxford: Oxford University Press, 2011). Here I mention Molinism only as an instance of a theory which violates the grounding principle and entails that choices can be free although externally, non-causally necessitated.

[24] Alfred J. Freddoso, *On Divine Foreknowledge* (Ithaca, NY: Cornell University Press, 1988)—this work contains extensive translations of Molina's texts; Thomas P. Flint, *Divine Providence: The Molinist Account* (Ithaca, NY: Cornell University Press, 1998).

[25] Rogers (2008): 148–51.

choice. There is literally nothing that could possibly ground such truth.[26] To my knowledge, outside of the supposed theological advantages of the theory, there is only one reason to defend the thought that there can be true counterfactuals of freedom, and that is that it helps makes sense of linguistic usages like, "If he had known it would become public, he never would have chosen to do that!" Granted, this is a common sort of locution, and one that we might well judge to have a truth value. But surely there is a simpler way to analyze this sentence than to posit a deeply mysterious realm of counterfactuals of freedom. Better to say that propositions which bear truth values and which are expressed through counterfactual sentences are statements about the actual world. Suppose you judge the sentence above to be true. What you believe, and what the sentence means, is probably something like: "He is (actually) the kind of guy who cares deeply about his public image, and now (actually) is ashamed or at the least very upset that this became public." It is about the actual world. And similarly, if you judge the proposition to be false, you think that "he" is (actually) the kind of guy who is so addicted to "that" sort of thing, that he can't refrain, even when he fears public disclosure and censure.

Anselm himself proposes a simple and elegant description of what the truth of a sentence consists in. A sentence is true when it says that what is, is, or when it says that what is not, is not.[27] Truth values are dependent upon the actual world. Counterfactuals or claims about possible worlds, fictional objects, etc. insofar as they admit of truth values, are about the actual world one way or another. On isotemporalism, truths about what is past or future relative to some perceiver are grounded in the actual facts of the matter. A presentist—that is, someone who thinks that only the present instant is real—has a little more trouble with truths about the past and the future. A presentist who is a universal determinist might say that what look to be claims about the non-existent past or non-existent future are actually claims about the present, in that the present entails everything about the past and the future. A presentist who is not a universal determinist might hold that what look to be claims about the past are actually claims about present "traces" left on the present by the past; evidence, memories, and such like. And what look to be claims about the future are actually claims about present anticipations of what is likely. (In that isotemporalism allows a more elegant analysis of propositions about the past and the future that might constitute yet

[26] William Hasker, *God, Time, and Knowledge* (Ithaca, NY: Cornell University Press, 1989): 18–52.

[27] *De Veritate* 2. He notes that any meaningful sentence has a sort of "truth"—it's doing its job just by signifying. His use of the term "truth" in this dialogue is very much broader than our usual usage.

another reason to adopt it.) The presentist who is a libertarian might do best to deny that there can be a present truth value to propositions about future free choices.[28] In any case, propositions which are expressed in a counterfactual form can be plausibly analyzed to be about the actual world. So the point that there are counterfactual sentences which look to have a truth value should not drive us to allow the Molinist claim that there are true counterfactuals of freedom which state how any possible agent would freely choose in any possible situation. If a free choice is an *a se* choice, then it is incoherent to suppose that there is a truth concerning the choice independent of the actual choice itself.

The Anselmian libertarian has a further problem with Molinism. Or perhaps this is another way of stating the problem. The claim was that, if a choice is an *a se* free choice, there is not a Molinist counterfactual of freedom about it. Another way to put the issue is: If there are Molinist counterfactuals of freedom, then the "free" choices which they are about are not *a se*, they are not free on the Anselmian understanding. The Molinist says that the counterfactuals of freedom are about libertarian free choices. But he is clear that the truth of the counterfactuals does not depend on the actual choices of actual agents.[29] How could it, since the counterfactuals exist independently of whether or not the agents who are their subjects are ever made actual? But now it is unclear just what sort of libertarianism is being espoused. I noted above that the classical theist must reject Molinism since the counterfactuals of freedom infringe upon divine omnipotence. They exist independently of God and circumscribe what He can do. Counterfactuals do not have *causal* power, but the limitation they impose is real, and readily admitted by (contemporary) Molinists. True, the Anselmian holds that God imposes a sort of limitation on Himself if He chooses to create agents who can make *a se* choices. The limitation is that He cannot simultaneously leave them free and control them, and that means that if He leaves them

[28] This is the conclusion that William Hasker (1989) embraces. But doesn't this mean that God is not omniscient? No, says Hasker, God knows all there is to know, but it is just impossible to know about future free choices. This is not a new approach, having been popular enough in thirteenth-century Paris to elicit condemnation from Bishop Tempier in the Condemnation of 1277 (#15). William of Ockham takes a somewhat different tack. He is apparently a presentist and a libertarian. He proposes that God presently knows truths about future free choices, and just admits that he doesn't know how that is possible; William of Ockham, *Ordinatio* d.38, q.u., M (Appendix 1, p. 90) in *Predestination, God's Foreknowledge, and Future Contingents* edited and translated by Marilyn McCord Adams and Norman Kretzmann (New York: Appleton-Century-Crofts, 1969). Ockham's proposed solution to the freedom/foreknowledge debate got quite a bit of play towards the end of the twentieth century, but the discussion seems to have moved on. For some fairly recent work on the Ockhamist approach see the related articles in *God, Foreknowledge, and Freedom* edited by John Martin Fischer (Stanford, CA: Stanford University Press, 1989).

[29] Flint (1998): 123–4.

free, they may choose badly, and He may not get the outcome He would prefer. But on the Anselmian account, this limitation is a *self*-limitation, and so it does not undermine divine sovereignty. The limitation which the Molinist proposes is imposed on God "from the outside." It is a brute fact about the universe, independent of God's nature and His will, that "In Q, S chooses B at t2," and God just has to live with it. But this is impossible on classical theism, and thus is sufficient for the Anselmian to reject Molinism.

More importantly for our purposes here, the counterfactuals of freedom also limit the individual created agent. The Molinist claim is that there have always existed true propositions about libertarian free choices. For example, the proposition "In Q, S chooses B at t2" has always existed. The truth of the proposition is not grounded in S's actually choosing B at t2. But S cannot do other than choose B at t2. If even God's activity can be limited by the existence of the counterfactuals of freedom then it is hard to see how the Molinist is to escape the charge that the "free" agent's activity is limited. I understand a determined choice to be one which is causally or ENC necessitated by factors originating outside of the agent. The Molinist will argue that the choice which is the subject of a counterfactual of freedom is not *caused* by the counterfactual. Granted. Nevertheless, the choice is ENC necessitated in the sense that, given the counterfactual, the agent cannot actually choose in a way which would falsify the counterfactual. If there is a counterfactual which says that "In situation Q, S chooses B at t2," then, if Q is actual, S "must" choose B at t2. It is necessary that S choose B at t2. And, though this ENC necessity does not impose a *causal* limitation on what S can choose, it imposes a real limitation. And the crucial point is that, unlike the consequent and the ontological necessities that the Anselmian allows, this ENC necessity does not originate with the actual choice of S for B. It is not a "necessity" which S imposes on himself simply by choosing. It is a necessity imposed by the existence of the counterfactual, a proposition existing independently of the agent. What distinguishes ENC necessity from the consequent and ontological necessities that are innocuous vis-à-vis free will is precisely that it is an "external" necessity. How, one might ask, does the counterfactual impose the necessity, if not causally? That is one for the Molinist to answer. ENC necessity conflicts with aseity, so there are no Molinist counterfactuals of freedom about *Anselmian* free choices. The Anselmian's commitment to aseity entails that knowledge of an *a se* choice, and the truth of propositions about an *a se* choice, must be grounded in the choice itself. The grounding principle will play an important role in Chapter Six, and the point that a choice that is ENC necessary cannot be an Anselmian free choice will be invoked in discussing the luck problem in Chapter Seven.

Character Creation

> The bad angel makes himself unjust and the good makes himself just.... From [God] they received both to have and to be able to hold on to or desert [justice]. God gave them the latter [the ability to desert justice], so that they would be able, in some sense, to give justice to themselves.
>
> *De casu diaboli* 18

A third entailment of Anselmian libertarianism grows out of the fundamental motive for developing the theory, the very reason that we want free will. As I explained in the introductory chapter, while the connections between freedom, responsibility, praise and blame, and reward and punishment, are all very important in the Anselmian schema, the core value in having freedom is that it renders the human agent a metaphysically elevated sort of being—something really special in the universe. Not only are we rational beings, which is certainly quite splendid, but we can also do something remarkable which other animals cannot, and that is we can make ourselves better on our own. We can act from ourselves—*really* from ourselves—and so contribute to our own creation. We do this by making *a se* choices, which choices constitute the building blocks of our characters.[30] Or, better to say, *some* of the building blocks of our characters. As the discussion of the Anselmian theory in Chapter Three should make clear, the Anselmian grants that the aseity of a created agent is quite limited. We do not bring ourselves into being. As small children we do not make *a se* choices. (The question of when we start to do so is interesting and open. More on this in Chapter Five.) Biology, heredity, environment, etc. (and God, according to Anselmian theism) supply us with our original characters. It is these original characters, and the situations in which we find ourselves early on, which set up the conditions for our first *a se* choices. In order to make an *a se* choice, one must find oneself in the torn condition, that is, one must want to pursue mutually exclusive, morally significant desires. And what desires arise to produce this condition will, at first, not be up to the young agent. But as we make more *a se* choices, we increase our autonomy by making more and more of our character our own.[31] We are focusing here on our moral character, so

[30] For some evidence from psychology that we can, indeed, deliberately improve our characters see M. Muraven, R. Baumeister, and D. Tice, "Longitudinal Improvement of Self-regulation through Practice: Building Self-control Strength through Repeated Exercise." *The Journal of Social Psychology* 139 (1999): 446–57.

[31] There is a theory, Situationism, which denies that we have characters, in the sense of established moral traits that express themselves in consistent behavior. And there are interesting experiments which purport to provide evidence for Situationism; see John M. Doris, *Lack of Character: Personality and Moral Behavior* (Cambridge: Cambridge University Press, 2002). The experiments, though, are amenable to alternative interpretations.

what is of interest are our virtues and vices. Included in character will be our conscious commitments to values.[32]

The Anselmian claim, going back to Aristotle, is that our choices and actions produce our character—they inculcate habits or propensities to behave in certain ways, virtues and vices, and—and this is what is most important for Anselm himself—a fixed internal orientation towards the good or the bad. One might have a sort of mixed character, in that regarding some character traits one might be properly oriented, while in others one might be wrongly oriented. For example, someone might be a glutton and yet be kind and loving to friends and family. I deliberately choose an "eating" example since it is useful to remember, as I noted in Chapter Three, that on the Anselmian view it is not the case that gluttony is to be considered a prudential, *as opposed to* a moral, issue. I am assuming the Anselmian approach in which there is very little that the agent might desire that is actually morally neutral and hence irrelevant to character building. Flourishing in the Aristotelian sense is a part of the moral landscape for the Anselmian. For Anselm himself, as discussed in Chapter Three, the torn condition arises when the agent desires justice on the one hand and, on the other, some conflicting "mere" benefit which is inappropriate. But the desire for justice is a second-order desire about how to regulate our first-order desires. We are always motivated by what we believe will be of some benefit for us—"benefit" very broadly construed. So eating is part of the moral landscape. Acting in a way that conduces to your health, and gratefully and appropriately enjoying the (for Anselm, God-given) goods of this world, are virtuous.

There might be some choices that are morally neutral, and hence could not constitute *a se* choice as described in Chapter Three. If you are debating between chocolate and pistachio then, *ceteris paribus*, you might feel "torn," but your decision is likely to be a mental coin flip and nothing like the struggle to pursue the morally better or worse desire. But your decision to have the ice cream cone right now is likely to have some moral dimension on the broad understanding of morality which I am assuming.

How does making the *a se* choice contribute to your character? Anselm's idealized example of the fall of Satan provides an extreme instance. That is why Anselm chose it. In the tradition of the classical theist brand of Christian philosophy the most puzzling instances of bad choices are those of Satan and Adam and Eve. They are puzzling because God is said to be perfectly good—

[32] Dean Zimmerman notes that libertarians often connect free choice and character creation. He gives libertarian theories that include this connection the apt label, "virtue libertarianism." See "An Anti-Molinist Replies." In Perszyk (2011): 163–86, at 176–7.

indeed His very nature is the standard for good—and all powerful and all knowing. He doesn't make anything that is naturally bad. The bad is a failure or absence or corruption or falling away from what is good. On classical theism it is just incoherent to posit something naturally bad. So God made the first rational agents good—both metaphysically and morally. And intelligent. And He made clear to them what His commands were. They were not subject to all the historical factors, especially all the evil and ignorance, which goes a long way to explain why humans now do what they do. So why in the world would a rational agent made good and intelligent and knowledgable choose to do something bad? Focusing on this most puzzling instance of choice allows Anselm to set aside the myriad details of the choices which constitute human history and pare choice down to its essence. Given his aims, this is a significant theoretical advantage. The obvious disadvantage is that Satan's choice and its consequences will not look much like what you and I and our neighbors are up to. Still, we can start with how S (here Satan, our paradigmatic created agent) formed his character and go from there.

S's choice between good and evil is unimaginably radical. Literally. ("Radical" from the Latin *radix*, meaning "root.") Anselm explicitly says that he will not venture to guess what inappropriate benefit might have motivated S to abandon his desire for justice. In being torn between this "mere" benefit and justice, S confronted the starkest choice possible. And he was intelligent and knowledgable. In this idealized instance there is no hint of an excuse for S's choice. And so, when S abandons justice, throws it away when he could have kept it, he plunges himself into a miserable condition. Choosing as he did constitutes turning against God. And one who so deliberately rejects the standard for good in the universe throws his internal ability to desire out of order. S cannot recover that desire for justice which he threw away. Now he can only will badly and stupidly. In making one and only one choice he has destroyed his character, and is now irrevocably fixed in vice.[33] (He is still metaphysically a good thing, on classical theism, since anything that exists at all, just insofar as it exists, is good. Why doesn't God do something to restore him? Anselm has an answer, but that is matter for a different study.)[34]

And for the angels who were torn, but clung to their desire for justice, in seeing the consequences for the bad angels, they now recognize that happiness lies all on the side of maintaining justice, and so they are unable to desire any inappropriate benefits. In holding on to justice when they could have thrown it away, they have irrevocably fixed their virtuous characters.[35] (Had the bad angels not fallen, God

[33] DCD 17 and 18. [34] *Cur Deus Homo* 2.21. [35] DCD 25.

would have presented the whole heavenly host with the knowledge of the consequences of evil in some other way. Anselm holds that God does not ever *need* some sin in order to produce a good He has in mind.[36] Those—like Augustine and Aquinas—who argue that God is the ultimate cause of the act of sin make this move, but Anselm explicitly rejects it.)

Is it just absurd to suppose that a single choice might change one's character radically and fix it forever? There are two questions here, one about changing one's character and one about fixing it. In the idealized example of S, his character is changed. Anselm does not see S as neutral in the beginning, but rather as good. S, after all, desires justice. S is not as good as he could make himself by freely clinging to the desire for justice when he could throw it away by per-willing the mere benefit. So S changes his character as well as fixing it. (There is the additional point that the desires in question in this case are all from God. S has had no time to make some *a se* choices which might have made him somewhat responsible for the desirable options he confronts in the torn condition. The suggestion might be made that S is just not responsible at all since he is not responsible for being in the torn condition. This issue is the subject of Chapter Five.)

I take it that, in the reality we know, and assuming that our free choices do build our moral characters, how this happens is not a purely theoretical issue. It is also, indeed primarily, an empirical question. My impression is that currently the disciplines that might study this from the perspective of empirical science are skeptical about free will, objective morality, and moral character (in the robust way I am understanding it), and so we will probably have to wait for the empirical evidence, if indeed it is ever forthcoming.[37] In terms of developing the theory we must, for the time being, be satisfied with asking about what seems likely and what seems possible. We can appeal to introspection, anecdote, and artistic representation of the human condition. And these are helpful sources. Here I hope only to give an outline of how such a discussion might proceed.

It seems to me, based on the evidence of introspection, etc. that the more likely procedure for character building is incremental. If we do make *a se* choices, we probably begin to make our first *a se* choices as children or young teenagers. We do not need to settle exactly when we begin in order to spell out the theory. Some

[36] DCD 25.

[37] Perhaps reticence concerning questions which relate to morality and virtue and vice relevant to the human condition would demonstrate an appropriate humility on the part of social scientists and neuroscientists. But on the other hand, it might be possible to generate useful working concepts of these moral entities such that they could be studied empirically. Here would be an opening for philosophers to engage in that "interdisciplinary" research so dear to academic administrators.

ability to recognize your moral choices and reasonably assess them is a *sine qua non* of *a se* choice. Chances are different people come to have this ability, and to find themselves in a situation which produces the torn condition, at different ages. It seems plausible to hold that children are much less responsible for their early *a se* choices than normal adults. This would follow from having no responsibility for one's character before one has begun to make *a se* choices, and from the mitigating factor that children have a less well-developed ability to recognize options, and to reason about them, than normal adults do.

Peter van Inwagen casts doubt on the thesis that certain mitigating factors result in degrees of responsibility.[38] And the Anselmian agrees that, considered in one way, responsibility is a you-either-have-it-or-you-don't phenomenon. If you do not make *a se* choices, then you do not have the sort of moral responsibility which is of interest. Anyone capable of *a se* choices has what is necessary for what we can call "basic" responsibility. But surely we want to allow that there are degrees in regard to how responsible you are for your choices and your deeds. (Some might prefer to say that the degrees attach to how praiseworthy or blameworthy you are. I don't know that it makes a lot of difference. I will stick with the locution that holds that there are degrees of responsibility.) Intuitively, it seems absolutely correct that we ("we" being society and the law) should hold children, and those with mental disabilities or terrible upbringings, less responsible for their choices and actions than we do normal adults.

In terms of character building it seems likely that children are more malleable than adults. As we grow older we become more "set in our ways." This means that, while children may begin to form their characters by *a se* choices, the characters are less fixed than they may become later on. Moreover, early on, it seems likely that one's choices tend to be about smallish things. Children, one supposes, do not have the same scope for moral choice that adults do. And it seems plausible to suppose that an *a se* choice about a small thing will have less impact on your character than a choice about something of enormous moral significance. So the picture that is emerging here is one where the project of self-creation begins tentatively, with small steps, which, chances are, can be retraced. So, for example, if a parent notices that a child is beginning to develop the habits of gluttony, the parent can encourage the child to exercise restraint and, one

[38] Peter van Inwagen, "Genes, Statistics, and Desert." In *Genetics and Criminal Behavior* edited by David Wasserman and Robert Wachbroit (Cambridge: Cambridge University Press, 2001): 225–42.

hopes, those bad habits can be undone and the project of developing the virtue of self-disciplined eating can be instigated.[39]

Continuing into adulthood, the project of self-creation seems likely to be a slow process, a constant interaction between the agent and the world, taking the form of incremental steps in building, and perhaps tearing down and rebuilding, various aspects of one's character. Changes in character are likely to occur so slowly that they are imperceptible, unless you view the agent over a long period of time and compare the agent at widely separate temporal intervals. This may be true even for the agent as he looks at himself, and even if he knows that he is trying to "work on" some aspect of his character. (But again, this is an empirical point. Sometimes one can see a relatively quick change. Some anecdotal evidence: I know a woman who gave up malicious gossip for Lent one year, about forty years ago, and she has been less inclined towards being a malicious gossip ever since.)

The usual path in character creation is probably long and incremental. But this does not rule out the possibility that sometimes—as in Anselm's paradigm case of Satan—a single choice may change you radically. In some lives that we know about great and rapid changes occur. Imagine—and this really is just an imagining, but one which I hope does not sound obviously absurd—Thomas Becket after Henry II has appointed him Archbishop of Canterbury. He has always been the King's man, interested in having power and a good time. He has always been a good politician but not a virtuous character. But he is familiar with stories of the great saint who was Archbishop of Canterbury earlier in the twelfth century, St. Anselm. Imagine him unable to sleep, struggling between continuing in his old way of life and repudiating it to follow the path of his saintly predecessor. We know that, in fact, Becket chose the latter course, much to everyone's surprise, Henry's not the least. It is not obviously absurd to suppose that he made an *a se* choice to abandon his former way of life and embrace the Anselmian modus vivendi; that he went to bed the King's man and arose committed to patterning his life as Archbishop after that of St. Anselm. (Note that the most important choice Becket makes, in this imagining, is not about some specific deed he intends to do, but about how he intends to pattern his future overall.) I do not say that this is what happened, but just that it is the sort of thing that could have happened.[40]

[39] As part of his discussion of the luck problem Alfred Mele (2006: 129–32) suggests a "soft libertarianism" in which "little agents" make non-determined choices, where the choices are relatively trivial and so the "lucky" element isn't so problematic. If these early choices do help to form the character of the agent, it is difficult to see why the triviality of the choice mitigates the luck problem. The luck problem is the subject of Chapters Seven and Eight.

[40] I claim no knowledge at all of the details of Becket's life and conversion. Perhaps many historians would want to resist the historical possibility of this imagining based on the methodology of the discipline where a goal is to try to explain people's behavior. In my imagining, there is no

In this story about Becket, not only does his character change, but it remains fixed in the sense that he continues to walk the Anselmian path until he is murdered for doing so. (Anselm, too, had trouble with the English king, but he was exiled rather than killed.) In Anselm's discussion of Satan, a single choice fixes *irrevocably* the character the angels have produced in themselves for good or ill. Is that possible in some human lives? One makes a single choice which sets one's character immutably? It seems unlikely, but not impossible. In the Anselmian wager in Chapter One I hypothesized that you have a friend who is a recovering glutton or alcoholic. He's been "clean" for eight months and now desperately wants that cupcake or drink. It is not obviously absurd that if he resists this time, it will be just the little extra bit of self-discipline added to the moderate habit he is trying to develop, that renders future temptations too weak to gain a purchase. Conversely, abandoning his efforts to resist might plunge him back into his bad old addictive ways such that he never recovers. I said above that "small" choices are likely to have only a small effect on one's character. But here what look to be small choices have huge consequences. Whether we think of such choices as genuinely "small" but nonetheless determinative—"the straw that broke the camel's back"—or as actually "large" and momentous choices, doesn't really matter. The point here is that it is not obviously absurd to suggest that, just possibly, a single choice might "fix" one's character. Again, it is an empirical question. (More on this issue in Chapter Five.)

A final point about choices, character, and the relationship of character to well-being. The point can be approached by considering briefly one of the main issues in the free will debate, the justification and nature of punishment. I will have more to say about punishment in Chapter Eight in the context of discussing the luck problem, and what it is exactly that we hold you responsible for. Here I want to spell out one particular aspect of punishment that is relevant to the Anselmian understanding of the subsequent effects of choice on the free agent. The Anselmian addresses the question of punishment from a perspective that used to be quite common, but is unusual in the contemporary free will literature. It deserves more of a voice. Today the picture that is often presented regarding punishment is that some agent, S for example, does some blameworthy deed, and then our question is whether or not it is appropriate to visit some pain or suffering upon him in response. And then, the standard modern approach would conclude, if

problem for the Anselmian if the historian offers a complete explanation for how Becket comes to be in the torn condition. But if the historian insists that the conversion could not have come about the way I suggest—through an *a se* choice—since that would render it inexplicable in terms of Becket's history, then he begs the question.

"we" do not visit the pain or suffering upon him, S has gone unpunished. But that is not the way the Anselmian looks at it.

Take Anselm's example of Satan's fall. Someone might argue that it seems absurd—cruel and unjust—for God to visit unending harm on Satan for one bad choice. But this supposes that the harm and suffering visited upon Satan is extrinsic to Satan's choice and the character that that choice produces, as if Satan could make his evil choice and be fine, if God didn't step in and punish him. But that is not the Anselmian position. By rejecting God, Satan cuts himself off from the source of value and happiness in the universe. Satan *makes himself* into a creature incapable of anything but misery. His punishment is self-inflicted.

Anselm's ethics are eudaemonistic.[41] (I do not say that Anselmian libertarianism, as a theory about the mechanics of a free choice, requires this approach, but it is helpful to see how Anselm and his contemporary followers think about character creation, since that is a part of the overall system.) On this sort of ethics, the good choices and deeds are those that cannot help but conduce to the happiness of the agent himself. As Socrates and many a philosopher after him insisted, and Anselm takes for granted, the construction of a good character is the construction of a happy life. And, conversely, you bring misery upon yourself by behaving badly.[42]

Take your friend the recovering alcoholic again and, to make the example vivid, suppose that he has been living at your house. If he takes that one little drink and returns to being an active alcoholic, you may decide that you cannot have him in the house any longer. He has made himself into someone who is unbearable to be around—a burdensome, abusive, happiness-vampire—and for your own sake and that of your family, you have to tell him to leave. He may feel that you are visiting a cruel and unjust punishment on him. And all on account of one little drink! But it is literally true that he has brought the punishment on himself, not as something additional, unconnected to his condition, but as a consequence of what he has made himself into. Or take your friend the glutton. One might say that his being unhealthy and immobile and ugly is a punishment for his gluttony, but the punishment is obviously intrinsic to the sin. One of the reasons gluttony is wrong is that it makes the glutton unhealthy and immobile and ugly. (It also damages the glutton's relationships with others, which, since we are naturally social animals, is yet another source of pain and suffering for the glutton.)

[41] Anselm's eudaemonism looks different from Aristotle's in that Anselm is thinking in terms of everlasting happiness with God, not just a happy life which ends at death.

[42] Rogers (2008): 58.

The punishment that society visits upon the convicted wrongdoer may sometimes be extrinsic. That Bernie Madoff—who stole large investments from people he befriended—is now in prison is not the inevitable consequence of his being a thief. He might never have gotten caught. But from the Anselmian perspective—even setting aside the possibility of an afterlife, though that possibility adds a *significant* dimension—it's just a bad idea to be Bernie Madoff. Being greedy and selfish and dishonest and willing to steal from those you call your friends just isn't worth the price. Sure it's nice to have lots of stuff (I imagine), but being vicious in order to get it cuts you off from the deeper and more lasting satisfactions—like being able to hope that you will leave the world a somewhat better place than you found it. That, at any rate, is the way Anselm and Anselmians think about the consequences of bad behavior.

Finally, one more important part of the Anselmian picture; if your self-formed character should causally necessitate future choices, you are responsible for those choices.[43] They are *a se*, in the sense that they are *self*-determined. Here is yet another sort of necessity which is innocuous vis-à-vis the freedom of the agent. Like the consequent and the ontological necessity mentioned in section two of this chapter, on the assumption that the character was freely formed, the causal necessity producing the character-determined choice is one which the agent freely imposes upon himself by making *a se* choices.

In the recent literature on free will an interesting and important criticism has been leveled against this sort of view. It is known as the tracing problem, from the claim that your responsibility can be "traced" back through your character to the free choices which produced your character. There are two facets to the tracing problem, both of which have to do with your knowledge as you make the choices which build your character. The first is this: as you made the choices which built your character you could not foresee what choices would confront you in the future. But since you were ignorant of these future choices, and ignorance can be an exculpatory factor, can we truly blame you for the choices determined by the character, when you had no idea, as you built your character, that you would face those choices?[44] And here is an even more fundamental problem: We trace your responsibility for the present choice through your responsibility for your

[43] Recent versions of this position can be found in Peter van Inwagen, "When Is the Will Free?" In *Philosophical Perspectives, 3* edited by James Tomberlin (Atascadero, CA: Ridgeview Publishing, 1989): 399–42; Robert Kane, *The Significance of Free Will* (New York: Oxford University Press, 1996); Timothy O'Connor, "Degrees of Freedom." *Philosophical Explorations* 12 (2009): 119–25.

[44] See Manuel Vargas, "The Trouble with Tracing." In *Midwest Studies in Philosophy 26: Free Will and Moral Responsibility* edited by Peter A. French, Howard K. Wettstein, John Martin Fischer (Guest Editor) (Oxford: Blackwell Publishing, 2005): 269–91.

character. But if, as you make your character-forming choices, you don't understand *that you are forming your character*, it seems we should not hold you responsible for forming your character. If this tracing problem could not be solved, a significant pillar of the Anselmian system would be undermined. Happily there are responses which I believe to be adequate. There is quite a bit to say on the tracing problem and possible responses, so I devote Chapter Nine to it.

Having set out the basic outline of Anselmian libertarianism and its entailments, I attempt to respond, in Chapters Five through Eight, to a number of standard criticisms leveled against libertarianism. Sometimes it will be obvious that the Anselmian approach improves upon other attempts to defend libertarianism. For example, in Chapter Six, having God as our would-be controller, and the grounding principle front and center, makes it very clear that Frankfurt-style counterexamples cannot undermine Anselm's insistence on alternative possibilities. But sometimes Anselm's theory seems more susceptible to criticism than other versions of libertarianism. For example, it has been argued that, if the agent has no control over the factors which motivate his non-determined choice, he cannot be said to have any control worth wanting over the choice itself. Anselm's example of Satan, whose condition is entirely determined by God up until the actual moment of the fateful choice, makes this criticism very vivid. And the contemporary Anselmian cannot disavow the gist of the example. The parsimony she hopes to maintain depends upon locating the indeterminism at the very moment of choice, and that means allowing the usual causes—divine and/or natural—to operate, even deterministically, through the torn condition that precedes the choice. If Anselmian libertarianism is to be defended the criticism must be answered. I take up the task in the next chapter.

5

Defending Anselmian Internalism

It is because the angel is given both [the desire for rightness and the desire for benefit] that it can be just and happy.

De casu diaboli 14

Introduction

The quote above emphasizes two key theses of Anselmian libertarianism; one is the standard libertarian claim that the free agent must confront open options. The other is that these options are "given" to the agent. That is, the Anselmian allows that the torn condition which precedes an agent's *a se* choice might very well be the product of causes outside of the agent's control. This is brutally clear in Anselm's paradigm case of Satan. Satan, as Anselm has it, was recently created, and everything about him—before he engages in his damning free choice—is from God. But then a legitimate question arises: If, in making his first *a se* choices, all that is ultimately up to the agent is that brief moment of choice itself, is this really significant enough to ground any sort of freedom worth wanting? If the Anselmian allows that there can be cases where *nothing* in the build up to the choice comes from the agent, why should that brief moment make the difference between freedom and unfreedom?

I believe that Anselm's view can properly be called an "internalist" theory, but with significant qualification. By an internalist theory I mean that it locates the criteria for an *a se* choice within the structure of the choice itself. Two important qualifications are in order, though. First, the structure of the choice includes a little bit of history. As I emphasized in Chapter Three, *a se* choice requires that the rational agent be in what I have referred to as the "torn condition," struggling between two, mutually exclusive, morally significant options. The struggle ends when the agent opts for one over the other. But only this little bit of history is required. *How* the agent came to find himself in the torn condition may be quite relevant to the degree of autonomy and responsibility that we can assign to him, but for some minimal responsibility all that is required is that the agent be in the torn condition and then choose. A second qualification was spelled out in Chapter Four and will be defended in Chapter Nine, and that is that a character-determined choice may be free and responsible if the agent's character was produced through his past *a se* choices. So I do not say that every free and responsible choice must have the structure of an *a se* choice, but rather that all free and responsible choices are either *a se* choices or are "traceable" to *a se* choices. But in that the locus of aseity lies within the structure of the *a se* choice (preceded by the torn condition), it seems proper to understand Anselmian libertarianism as an internalist theory.

Recently internalism has come under attack, especially from Alfred Mele, and it is an attack which, at least prima facie, looks to have some traction against Anselm's version, so it is important to offer a defense. First I spell out internalism and externalism, as I will be understanding them, with some introductory remarks. Then I turn to Mele's criticism of internalism in the form of an example which he raises against Kane. I argue that his example does not capture the sort of choice the Anselmian has in mind. But the Anselmian is not in the clear. I look at a further criticism of internalism that Mele raises against Frankfurt, but which could also be aimed at the Anselmian, and argue that, if we fill in Mele's example a bit more, the intuition that he hopes to elicit is not likely to be forthcoming. Then I offer a new example which meets Mele's criteria for the sort of internalist situation which Mele holds to undermine responsibility and argue—or at least try to elicit the intuition—that our agent in my new example ought to be held responsible.

The examples Mele proposes to show that internalism is wrong-headed involve agents who are suddenly imbued with new values. I note that one versed in the Christian tradition may not find this sort of situation corrosive to responsibility in that he will be used to assuming that such conversions happen rather often and do not undermine responsibility. I conclude by allowing that the sorts of historical circumstances which Mele takes to ground responsibility do certainly *enhance* responsibility. And conversely, the Anselmian has to grant that someone with only *basic* responsibility may have severely limited autonomy. The adult whose character is fixed by childhood *a se* choices is at least a theoretical possibility. And there is the practical problem that, in assessing responsibility, the third party (other than God) suffers under epistemic ignorance concerning the agent's past. I spell these problems out and attempt to mitigate them somewhat.

Another important reason to spend a chapter on internalism is that the discussion of internalism is likely to bring to mind what is currently held to be the biggest difficulty with libertarianism, the luck problem. And so this chapter can help to set the stage for that issue. But the luck problem itself must be set aside until Chapters Seven and Eight.

Internalism and Externalism

The Anselmian focuses on aseity as the basic requirement for freedom and responsibility. We do not bring ourselves into being, so we must confront alternative possibilities in order to choose from ourselves. In that ultimate responsibility is grounded in the capacity for non-determined choice between open options, and not in the agent's history (except for character-determined choices), Anselmian libertarianism is what Alfred Mele labels an internalist theory. ("Structuralist" is another name for this sort of theory.) That is, the necessary and sufficient conditions for responsibility-grounding freedom are found within the act of choosing itself. The externalist (or historicist) holds, to the contrary, that it is the events and processes in the history of the agent that constitute the conditions for freedom. Mele argues against internalism, and so it is important for the Anselmian to respond.

Anselm's idealized example of the fall of Satan makes his internalist position vividly clear. Everything about Satan (S) before he makes his choice for good or ill is caused by God. How, then, can S (along with the rest of the hosts of heaven) have the ability to make himself better *on his own*? Only by it being absolutely up to S what he chooses, at the very time he makes his choice. Similarly for human agents, especially early in their careers as choosers; it may be the case that they have no control at all over the factors that have led to them finding themselves confronting open options, that is, to being in the torn condition (TC). So if aseity is what ultimately counts, it cannot be the agent's history that grounds his responsibility.

There is a prima facie advantage to Anselmian internalism (and perhaps this is true for most internalist theories) over externalism, at least over the sort of externalism Mele proposes in his *Autonomous Agents* (1995).[1] Here Mele proposes that the autonomous agent (Mele couches his proposal in terms of "autonomy" rather than "free will") must have a long and complex history. Or, if not such an actual history, at least the numerous, robust abilities that, in the usual course of things, could be generated only by such a history.[2] (I offer a brief sketch of his list of requirements in Chapter One.)[3] This means that many apparently

[1] Mele is a prolific writer and has a great deal to say about what is important in the histories of agents. In the present chapter I use Mele as a foil, but I do not claim to do credit to his many and varied discussions concerning the present issue. I hope to say just enough to spell out and defend Anselmian libertarianism.

[2] Mele (1995): 172–3. Here Mele allows the possibility of an autonomous agent who has just come into being with her abilities intact.

[3] Mele would, I believe, now grant that it is possible that even very young children bear some moral responsibility for their actions. See Mele (2006): 129–32. My criticism here is directed at the Mele of 1995 who proposes a long and complex list of attributes for autonomy.

normal adults are likely to be left out of the circle of autonomous agents. And this is a difficulty given the methodological presupposition about inclusiveness which I proposed in my introductory chapter. On the Anselmian understanding, it is demeaning to someone to hold that he is not a responsible agent. There should be resistance to setting the bar for free agency very high. The thought is that between two theories which are equally plausible in other respects, if one theory would include more of humanity within the fold of the morally responsible agents, it is the preferable theory. Theories that demand a long and multi-faceted history in order to ascribe free agency to the agent are exclusive and this is something of a strike against them. Anselmian internalism, which grounds basic responsibility in the ability to make *a se* choices would—assuming there *are* such things as *a se* choices—allow for the inclusion of more of humanity as morally responsible agents. But, of course, this prima facie advantage is worth little if the theory cannot be plausibly developed and defended.

Opponents of internalism argue that—even assuming the agent is rational and struggling between options which he recognizes as mutually exclusive—the minimal ability to choose otherwise is so anemic that it is neither sufficient nor necessary for responsible agency. Mele makes this case against Robert Kane's libertarianism.[4] (And also against Frankfurt's compatibilism as discussed below.) Kane's libertarianism, as I noted in Chapter Three, is significantly different from Anselm's. Anselm takes the free choice itself to be a non-determined event, where what precedes it may be determined, though the preference for one option over the other is absolutely up to the agent. Kane describes the free choice as the terminus of an indeterminate process in which one or the other of the competing motives (or "efforts") causes the choice probabilistically. Kane's description seems more susceptible to Mele's criticism than does Anselm's, in that, for Kane, it is a preceding motive that causes the choice. (Anselm allows that the motive power comes from the preceding desire, but the *opting* for this over that, comes from the agent.) And so, on Kane's understanding, if the agent bears no responsibility for his having or pursuing the motive, it seems problematic that he should bear responsibility for the choice it causes. The mere indeterminism of the preceding process does not seem to contribute anything that would respond to this difficulty. (This is essentially the standard critique of event-causation noted in Chapter Three.)

Anselmian libertarianism is quite different from Kane's, nevertheless it is still close enough in some respects that it might seem to fall prey to Mele's criticism of internalism. After all, if everything about you before you choose is determined,

[4] Mele (2006): 52.

why should that tiny slice of aseity in your life make such a difference? I will defend Anselmian libertarianism against Mele's externalist (or historicist) criticism, adopting the artifice of responding as if Mele had leveled his criticism against the Anselmian version of a libertarian free choice. Of course, I do not insist that Mele himself would propose just the same criticism, expressed just the same way, against Anselmian libertarianism.

In the end I grant that Mele's criticism gains enough traction against the theory I am developing that I must admit that the agent who is free and responsible by Anselmian criteria may, in many instances, bear only very limited responsibility. I do not dispute the claim that there is great value for the agent in having the sort of history which Mele holds to be required for autonomy. I argue only that without aseity, the history does not provide for robust responsibility and the ability to self-create. (Let "history" stand for the history and/or the relevant abilities, to take account of the agent magically just created with the autonomy-making abilities.) And, even without the history, an agent who chooses *a se* has at least basic responsibility.

Antti: Mele's Critique of Libertarian Internalism

Mele proposes an intuitive argument through an example which, he takes it, fits the description of a Kane-type free choice. He believes that the example should evoke the intuition that the hypothesized agent is not free. He asks us to "imagine that a manipulator compels an agent, Antti, simultaneously to try to choose to A and to try to choose to B, where A and B are competing courses of action that, in the absence of manipulation, Antti would abhor performing." Antti cannot choose anything other than A or B, and upon choosing one or the other will endorse the outcome as something he was wanting or trying to do. "The tryings are internally indeterministic, but Antti does not freely try to make the choices he tries to make." And so, Mele concludes, "Apparently, whatever he chooses, he does not freely choose it—especially when the sort of freedom at issue is the sort most closely associated with moral responsibility."[5] But suppose the example is developed so that it is clearly an instance of *Anselmian* choice. In that case it is doubtful that Mele's intuition would be widely shared.

Note first that as the example is set out, it seems conceptually possible that the manipulator might compel Antti to try to choose A or to try to choose B without providing the background motives—the desires, intentions, etc.—which Anselmian (and Kanian, it should be noted) free choice involves. The manipulator

[5] Mele (2006): 52.

might have compelled Antti to engage in a sort of brute, irrational, and unmotivated "trying," which just is not the sort of struggle to pursue conflicting motives that the Anselmian envisions. And the locution that the "tryings are internally indeterministic" is unclear. One way of reading it echoes the way in which Mele sets up the luck problem for the libertarian. Antti in the example seems passive, almost as if the "tryings" were something the manipulator is doing through him. Mele's description of libertarian choice implies that it is "just a matter of luck" that the agent finds himself in the actual world where he chooses one option, rather than in an alternate possible world in which he chooses the other option.[6] But the Anselmian insists that it is the agent's own actions that make the actual world to be the actual world. So the agent's being in the actual world is not, as Mele's description suggests, something that happened to him. (See Chapter Seven.) To come closer to Anselmian choice, assume that the manipulator has provided Antti with motivations towards A and B. And take it that "the tryings are internally indeterministic" means that Antti actively struggles to pursue each one, where it is not determined which he will succeed in choosing. These qualifications bring the example closer to being one of Anselmian choice.

But there is still an important difference. Mele fails to include one of the important criteria for free and responsible Anselmian choice; the choice between A and B must have moral significance.[7] In ordinary circumstances, the choice between coconut and pistachio ice cream, for example, won't constitute an Anselmian choice. Mele says that both A and B are options that Antti would abhor, absent the manipulation by the hypothetical controller. Presumably manipulated-Antti does not find them abhorrent now, since, by hypothesis, he desires to do them. Does he consider them morally important at all? If so, given that before the manipulation they were both abhorrent, does he still find them "equal" in some morally relevant way, so there is no moral significance in choosing? If Antti's options present a choice which he does not find morally significant, then the choice here does not fit the description of an Anselmian free choice. Whether Antti chooses A or B, he cannot be held morally responsible for making the better or the worse choice if he does not see a better or worse.

Moreover, Mele clouds the issue by having the manipulator impose the pursuit of out-of-character motives on Antti (assuming we can understand the "tryings" not as brute, unmotivated acts, but as Kane and Anselm understand them, the pursuit of rational desires or motives). This introduces an identity problem. Has

[6] Mele (2006): 51.

[7] Kane might add that other choices can be self-forming. So, we might say that Antti needs to confront moral options, or options that have a similar significance in terms of self-formation.

Antti over the years been making free choices to develop a certain character, and now the manipulator erases that character and replaces it with new traits? Is Antti, then, the same person he always was, just with a few new, odd, and anomalous interests? Has Antti been radically changed? Has he ceased altogether to be, and his place been taken by a new person? Mele's example poses identity questions which distract from the main issue, the freedom of the agent. And Mele does not mention the question of Antti's subsequent evaluation of the choice. According to Mele, Antti understands it as *his* choice—a choice he was wanting or trying to make—but afterwards does Antti embrace or reject it? Given Antti's very odd circumstances, his subsequent adoption of a certain attitude towards his choice may have more moral significance than the choice itself.

Suppose that Antti is in fact confronted with a morally significant choice and that the correct way to describe Antti's identity is that he is still truly Antti, but with some new interests and values. If it is indeed up to Antti which of the motives in question to per-will, then the Anselmian maintains that Antti is free and morally responsible. The conclusion would be the same if the manipulator had succeeded in exterminating Antti and so had replaced him with a new person. Or if, as in Anselm's own example, the manipulator—God—had brought Antti into being in the torn condition, struggling to opt for either an inappropriate benefit or justice.

What Mele's example suggests is that, on the Anselmian theory, responsibility, thought of as praiseworthiness or blameworthiness, admits of degrees. Assuming that Antti is a normal human adult (Mele's description implies that Antti has had a history as an agent, so he seems to be an adult), and does not have the power and knowledge of a newly made angel, then Antti's responsibility for his choice is limited by the fact that the options were imposed on him. In this he is rather like a child. But it does not follow that he has *no* responsibility. He has a grain of aseity in that the choice was up to him, and so he can be praised or blamed a little bit. And if he is free to make a subsequent choice to embrace or reject the previous choice, then he may be on his way to developing his character on his own, in spite of the role of the manipulator. I defend the Anselmian approach below, with a return to the example of Antti, but a quick sketch of another of Mele's examples helps set the stage.

Chuck and Beth: Mele's Defense of Compatibilist Externalism

Mele, in defending historicist compatibilism, offers another example—Chuck and Beth—which is similar in some respects to the Antti case. Examining this example will, I think, help make the intuitive case that, while Antti (adjusted to fit the Anselmian description of free choice) has a severely limited freedom, he is nevertheless a more appropriate candidate for praise and blame than one of Mele's compatibilist agents. It is having options, not history, that grounds responsibility. Mele defends externalist (or historicist) compatibilism against internalist (or structuralist) compatibilism. Harry Frankfurt, an internalist and a compatibilist, is Mele's specific target in the Chuck and Beth discussion. Frankfurt holds that "If someone does something because he wants to do it, and if he has no reservations about that desire but is wholeheartedly behind it, then—so far as his moral responsibility for doing it is concerned—it really does not matter how he got that way." He adds that, "the person's desires and attitudes have to be relatively well integrated into his general psychic condition."[8] But it makes no difference how the agent came to be as he is. (So, although Anselm disagrees with the compatibilism, the relevant point of agreement between Frankfurt and Anselm is that *how* the agent came to be in a position to make a choice is not what counts for an ascription of basic responsibility.)

Mele has described internalism as, "A view of agents as thin 'time-slice' entities."[9] That is a mischaracterization in that, even on internalism, whether Anselm's or Frankfurt's, agents must have a *little* history. As noted above, the Anselmian choice is the culmination of a temporal process involving the struggle between the two competing motivations. Frankfurt's free agents must, at the second-order level, take time to assess and identify with their first-order desires. But only a little history is absolutely necessary for the Anselmian or the Frankfurtian free agent. If Antti had been created an hour before he made his choice, with the motives and desires the manipulator gave him (a set of false memories would help to make Antti seem to himself to be a normal human agent), the choice could still fit the description of an Anselmian or a Frankfurtian free choice.

Mele proposes the example of Chuck versus Beth in attempting to make the case, against Frankfurt's internalism, that a long and fairly eventful history is required for an agent to be autonomous. Chuck and Beth are both inhabitants of a determinist universe. Chuck, from the time he was a child, has nurtured, practiced, and endorsed a value system in which cruelty and brutality are prized.[10] He enjoys killing and endorses his murderous desires. Mele says

[8] Harry Frankfurt, "Reply to John Martin Fischer." In *Contours of Agency* edited by S. Buss and L. Overton (Cambridge: Cambridge University Press, 2002): 27–31, at 27.

[9] Mele (1995): 165.

[10] Mele provides a detailed and lengthy discussion of what he takes to be necessary for autonomy, but for our purposes here it is enough just to note some features of the sort of history which Mele takes to be required.

that he and Frankfurt can agree that, when Chuck commits a murder, he is responsible. Frankfurt can point to his endorsing his desire to kill, which desire is "well integrated into his general psychic condition." But if we look at the example of Beth, says Mele, we will see that the structure of choice is not sufficient to guarantee responsibility. The historical development is important as well. Beth has always been a sweet person, for whom the thought of killing another human being could not even arise as an option. But in the night the manipulators come and extract her previous value system and replace it with Chuck's value system. They leave her memories intact, and she is somewhat surprised, upon awakening in the morning, to discover the reversal in values. But, now that she appreciates the plausibility of the nasty worldview, she embraces the desire to kill and commits a murder. Mele says that on Frankfurt's account, Beth should be held to be as responsible as Chuck. But his own intuition, which he believes would be commonly shared, is that it is obvious that Beth is not responsible.[11]

Mele says that the difference in intuitive responses to Chuck's and Beth's murders holds even if we add that both of them "could do otherwise" in some respectable compatibilist sense.[12] And it holds if the moral metamorphosis goes the other direction, from evil to good. If the manipulators infused Chuck with the original Beth-like values, he is not responsible for his subsequent good behavior.[13] And moreover, it is not the hypothesis of the manipulator that produces the difference in intuition. It should stay the same, according to Mele, even if we replace the manipulators in Beth's story with a natural phenomenon such as a brain tumor. If the brain tumor produces the reversal of values, Beth is not responsible.[14] The intuition that Beth is not responsible is ultimately driven, says Mele, by our noting that, before the transformation, her character—for which she is *ex hypothesi* morally responsible in some respectable compatibilist sense—was such that committing a murder would not have been an option for her.[15] At the time of their acts of murder Chuck and Beth are value twins. It is their history that is different. And so, if we recognize that Chuck, but not Beth, should be held responsible, it must be due to their having different histories.

But filling out the example a little shows Mele's intuition to be debatable, undermining his externalism. Let us (remembering the divine controller argument from Chapter One) say that God has caused everything about Chuck and Beth. Positing God makes the case vivid, but "God" here can stand for God, or for

[11] Alfred Mele, "Moral Responsibility and Agents' Histories." *Philosophical Studies* 142 (2009b): 161–81, at 166–8; see also Mele (2006): 171–2 and Mele (1995): 144–76.

[12] Mele (2009b): 168–9.

[13] Mele (2009b): 173.

[14] Mele (2009b): 168, fn.11.

[15] Mele (2009b): 169.

some non-divine manipulator(s)—perhaps we are characters in a video game being played by some super-human intelligence—or for natural processes. Mele said it should make no difference what produces the sudden value transformation, and my elaboration of the cases will be consistent with any of these three scenarios, *mutatis mutandis*. With God or the non-divine manipulator we can speak about their purposes, but on Mele's reading, whether or not the causes at work have purposes is irrelevant to the different intuitions he believes we ought to have regarding Chuck and Beth.

Say that God wants Chuck to kill Don. Knowing that if Chuck is in condition H at t (some particular time) he will kill Don, God brings it about that Chuck is in condition H at t. Preferring to work through secondary causes, He achieves His purpose in the following way:[16] First He causes Chuck to read *Beyond Good and Evil* at the age of thirteen. Call this reading "A." God then, following upon A, causes Chuck to experience and to do the series of thoughts, actions, and events which Mele describes; the nurturing and endorsing and choosing, which lead Chuck to have a value system in which he can feel no mercy. Call this series of thoughts etc. "B through G." God, with A–G as secondary causes, now causes Chuck's merciless condition, which includes his nasty values. These values are now unsheddable for Chuck. He cannot root them out of his thinking. But, Mele would have it, that does not infringe on Chuck's autonomy, since the values are the result of the sort of history that Mele believes to be autonomy producing. Call Chuck's merciless condition H. Finally God, with H as a secondary cause, causes Chuck to choose to kill Don and to kill him at t. Phenomenologically it looks to Chuck as if his killing Don was due to his being in condition H—that is his having the desires, values, etc. that led him to do the deed. And it looks to him as if he has inculcated these values in himself since his reading Nietzsche at the age of thirteen. And Chuck is right in his assessment of his history, on our hypothesis that God works through secondary causation. According to Mele, Chuck has the proper history, and so he is morally responsible for the killing, and he deserves to be blamed and punished.

Now Beth. God wants Beth to kill George. Knowing that if Beth is in condition V at t she will kill George, God brings it about that she is in condition V at t. V is the Beth-iteration of having the Chuck values. First God causes Beth to read *Pride*

[16] As noted in Chapter One this is how the God of classical theism, as represented by Thomas Aquinas, for example, goes about his business. It is right to say that fire burns cotton, but it is also right to say that it is God who causes the fire to burn the cotton, and hence the burning of the cotton, since God keeps the whole causal system in being from moment to moment. This analysis of divine causation makes it easy, for purposes of the Beth and Chuck examples, to replace God with natural processes.

and Prejudice at the age of thirteen, and to see the sweet and gentle Jane, rather than scrappy Elizabeth, as the heroine of the novel. Call this reading "O." Then He causes P–T, the thoughts, acts, experiences, choices, etc. which constitute Beth becoming the person whom Mele describes. She is thoroughly sweet, and holds and identifies with deeply nice values, such that she could not entertain the thought of committing a murder. God must cause one more series of events in order to bring it about that Beth becomes the value twin of Chuck. Last night Beth couldn't sleep. She has heard philosophy is a dull discipline, and her grandson, a philosopher, happens to have left a copy of *Beyond Good and Evil* on her bedside table. Under the impression that it should soon put her to sleep, she picks it up and begins reading. Without meaning to—she wanted to get to sleep—she becomes engrossed and keeps reading. After eight hours she is suddenly struck with the thought that her life up until now, modeled on Miss Austen's gentle Jane, has been one long exercise in the slave morality. She feels the will to power welling up within her. She is seized with an urge to become a master and assert her freedom through cruelty. Call this eight-hour series of events U. Now that God has caused Beth to go through O–U, she is in condition V, where she shares and endorses the same values as Chuck. God then causes her to choose, as her first overt master act, to kill George, which she does, this very day. Phenomenologically it looks to Beth as though she has, thanks to Herr Nietzsche, suddenly broken free of the oppression of the Judeo-Christian slave morality. Yesterday it would have been impossible for her to even consider killing George, but today she revels in her act of "throwing off the yoke." Her past "Jane" life looks so vapid to her now that she cannot even imagine going back. Condition V entails Beth holding the Chuck values in what Mele calls an unsheddable way.

Beth fits Mele's criteria for someone with the wrong etiology for responsibility. For years she was committed to a different value system in which she could not have wanted to kill George. She was responsible for that value system in that God had caused it in her through the process which, according to Mele, grounds autonomy. Her values were suddenly and recently transformed by a manipulator; God working through Nietzsche's opus, though a different manipulator, or a chemical imbalance in the brain, would have done as well. She did not consent to the divine activity, knowing nothing about it. And she did not consent to read the book, since she became engrossed without having made any commitment to keep reading.[17] And now she cannot (except possibly in some compatibilist sense and

[17] If continuing to read sounds too much like a consensual activity, hypothesize that she put on a disk her grandson had left, then fell asleep, so that someone is reading to her all night. The text affects her sleeping mental state such that her first thought upon awakening is the disgust at her past,

consistent with her new values) fail to do other than kill George.[18] Mele would have it that she is not responsible for killing George and does not deserve blame or punishment.

But this is puzzling. Isn't Beth's act caused in essentially the same way Chuck's is? In each case God wants to bring about a killing and produces, in Chuck and in Beth, the series of events, etc., which will cause them to kill. God causes Chuck to kill by causing A–H in him, and He causes Beth to kill by causing O–V in her. That is, God produced the same values with the same effect, the killing, just through two different series of causes. Mele can say that it is the difference between the two series that makes all the difference, but in that both are divinely caused and both entail living a life in which reading *Beyond Good and Evil* produces the Chuck values and the subsequent killing, it seems odd that so much hinges on *when* each agent read the book and just what divinely caused process produced the Nietzschean worldview and the murder. When Chuck and Beth come to trial they are both, let us suppose, unrepentant Nietzscheans. Would justice be served if the jury, having convicted Chuck, should acquit Beth on the grounds that she read Nietzsche only hours before she committed the crime? If that seems wrong, then Mele's intuitive support for his thesis that it is the history that counts is undermined. And if their differing histories do not ground a difference in their status as responsible agents, then it seems we should say either that they are both responsible or that neither is responsible.

And here is a further consideration. Suppose that God, wanting Beth to kill George, decides to effect this project this way: Beth goes to sleep her usual sweet self, and awakens next morning not only a Nietzschean of the Chuck variety, but with her former memories replaced by a new set. This morning she remembers a complete, Chuck-like history, from childhood on, involving all of the elements which Mele holds to be requisite for autonomy. At the time when she chooses to kill, phenomenologically from her perspective, her value system was produced in just the same way Chuck's was. Should Mele say that she is not responsible, in that she does not *actually* have the long Chuck-like history? But why? The past is gone for both Chuck and Beth.[19] Chuck and Beth are value twins in the present,

"Jane" existence. She then embraces the Chuck values. Mele does note that, "An agent who, in the absence of brain-washing and the like, sees the error of his ways and radically transforms himself might properly be credited for his own good behavior" (1995: 165). But the relevant distinction between brain-washing, other sorts of manipulation, and the causally determined "seeing" that Mele here takes to be innocuous is not clear.

[18] Mele (2009b): 171.

[19] The isotemporalist mentioned in Chapter Four, who holds all times to exist equally, might look at this point differently, but I do not suppose that Mele's historicist account includes that theory of time.

and phenomenologically, in terms of the relevant "facts" at present, they are history twins. Is there a difference in that it was, in some more robust way, *up to* Chuck to produce his present value commitments through his actual history? But in both cases it is God who is the cause of the present value commitments. With Chuck God accomplished this through causing in Chuck the actual historical events and processes producing the present, divinely caused, values and memories. With Beth, in this slightly altered example, God accomplished this more immediately by just causing in Beth the present values and memories. By hypothesis their two different histories do not produce any difference between Chuck and Beth in the present. Chuck and Beth have done the same deed and have the (relevantly) same values and memories. So it seems bizarre to convict Chuck and acquit Beth. Again, either they are both responsible or they are both not responsible. The Anselmian, of course, opts for the latter.

Antti Revisited

Given what we know about the causes of Beth's conversion, as I have described it, the Anselmian agrees with Mele's intuition to this extent—Beth is not responsible due to the divine manipulation. But then it seems correct to say that Chuck is not responsible, either. The Anselmian takes it that what the manipulation case shows is the importance of aseity. Neither Beth nor Chuck has aseity and neither should be blamed or punished for what God (or a non-divine manipulator or natural processes) made them do. To illustrate how aseity, rather than history, is what matters for responsibility, let us return to Antti. And now let us refer to him as Anselmian Antti, AA for short, to distinguish him from Mele's Antti. I believe we can flesh out the AA case to show that, although lacking the history that Mele takes to be necessary for responsible agency, AA, in the right circumstances, should be considered a responsible agent.

I situate AA in the real world in the hope of strengthening the intuition that AA is indeed a live possibility. Let us say that AA is a guard at Auschwitz. He read *Beyond Good and Evil* at the age of thirteen, and, over the years, *Mein Kampf*, some social Darwinist tracts, and assorted other instances of National Socialist propaganda. He has trained himself—through the sorts of processes which Mele attributes to autonomous agents—to be a committed Nazi. The thought of disobeying orders is disgusting to him. He cannot entertain it as an option. And now it is July of 1941. Three prisoners have escaped the camp. In retaliation, and to deter further escapes, ten more prisoners are sentenced to die. When one of them cries, "My wife! My children!" a Catholic priest and fellow prisoner, Fr. Maximilian Kolbe, volunteers to die in his stead.[20] God, wanting to offer Antti a chance to repent and be saved, takes this opportunity to bestow grace upon him. This grace takes the form of endowing AA with new values and insights. Upon witnessing the courage and the calmly endured sacrifice of Fr. Kolbe, Antti, by the grace of God, suddenly sees Nazi ideology as vicious nonsense and his former life as an absurdity of dust and ashes. He suddenly embraces a new set of values which includes repudiating his Nazi past and valuing human life in general. In the twinkling of an eye, what had been unimaginable the day before is a live option.

It is important, in trying to construct an example that responds to Mele, that the new values be inculcated in "the twinkling of an eye." Mele discusses examples of conversion. He grants that some traumatic occurrence, like getting hit on the head, or some other exceptional life event, might trigger a new insight which leads to new values. But for the agent to be responsible, he holds, the agent must, one way or another, exhibit the history Mele takes to be necessary for autonomy. So he suggests that perhaps the agent takes the time to consider and evaluate and reflect upon his new values. And then, upon consideration he finds them unsheddable and so, when he acts upon his "new" values, he does so responsibly, in that the history is there, although it was triggered by some sort of conversion experience. Or perhaps the values were already there, and properly formed, in the agent, and the apparent conversion is in reality an instance of some new insight gained which triggers some new way of looking at or expressing the already extant values.[21] But the case of AA is not like that. God has implanted in AA some genuinely new and unsheddable values. And, if the question is whether or not AA is responsible for his new values, the Anselmian answers with a resounding, "No! He is not!" God, and Fr. Kolbe—but mainly God—are responsible. But suppose, the instant after receiving his new values, AA turns to considering what he ought to do next.

It is too late to help Fr. Kolbe and the other nine victims of the retaliation. But perhaps there is a way for AA to pay for his past and throw in with that side of the universe that disapproves of piling up the bodies of the innocent dead. The thought leaps to his mind that he might join the Polish Underground. If his new values determine him to choose to do that, then he is not responsible for that choice. To be an instance of Anselmian choice, AA's situation must admit of open, morally significant options. If the divinely inspired conversion experience was such (as Augustine and Calvin might have it) that AA has no option but to

[20] Fr. Kolbe was canonized in 1982.

[21] Mele (2006): 179–84.

follow a fixed path to glory, then he is not really free.[22] So let us say that AA, though completely repudiating his past values, still confronts morally significant options. Now that he values human life, he suddenly feels a fear of death that he had never known before. Yesterday the idea of a young, Wagnerian death was appealing to him. But now he is sorely tempted to just sneak out of the camp, run away, and lie low in a distant relative's house for the duration. He recognizes this as the morally worse option. I take it that AA is in the situation Mele described. He has a new set of unsheddable values which were provided by a "manipulator." His possession of these values does not have the sort of history Mele believes is requisite. And his two options, which are rooted in these new values—lie low or join the Underground—are options which he would have found abhorrent only an instant before. He struggles with his two motivations and finally makes an *a se* choice to join the Polish Underground. Contra Mele, it seems entirely reasonable to consider AA responsible and to praise him for his decision.[23]

It is true that God and Fr. Kolbe had a big hand in leading AA to the point where his options are to run and hide or to join the Polish Underground. Very little was up to him. But in the Chuck and Beth stories *nothing* is ultimately up to them. It is *all* up to God. AA may not really be very praiseworthy, since it took divine intervention to dislodge him from his vicious path, and humility would be an appropriate attitude for him to adopt. But he seems at least a little responsible for his choice to join the Polish Underground. The Anselmian will say it is precisely because he could have chosen to lie low. If this conclusion is correct, then it is aseity, and not Mele's historical criteria, which grounds responsibility. The minimum requirement for an agent to have what I have called "basic" responsibility is aseity, the ability to make an Anselmian choice. That seems to me to be the conclusion which intuition arrives at when Mele's cases of Antti, Chuck, and Beth, are fleshed out in the ways I suggest.

Conversion Intuitions

But here, perhaps, is a point where intuitions may differ based on background belief. It is standard in Christian thought to hold that you are often not responsible for the situation in which you find yourself—including what gifts and talents you may have—and yet that you are robustly responsible for what you make of

[22] As Anselm himself spells out the mechanics of grace, the one who receives it can reject it, likely by returning to his bad old ways. See Rogers (2008): 136–41. To fit the criteria of Mele's Antti case, however, AA must find himself in a situation in which the old ways no longer have any appeal.

[23] It is true that AA acts "out of character," but the Anselmian praises him for what he is becoming due to his choice, not for what he was before his choice. See Chapter Eight.

what you have been given.[24] And the traditionally minded Christian philosopher is more or less stuck with accepting sudden conversion where genuinely new values are implanted in an agent willy-nilly, yet the agent is considered free and responsible. That is what grace consists in. Indeed the question of grace has driven the free will debate for much of its 1,500-plus-year history. Discussion about free will goes back further, of course, but Augustine turned up the heat by arguing that grace—which aims the human being, who is inevitably sinful, back towards God—is necessary for salvation, unmerited, and irresistible. As his opponents noted at the time, that seems to leave no room for human freedom. To which Augustine responded with a defense of compatibilism.[25]

The debate was still going in Anselm's day, but it was Anselm who realized that, in order to find a solution, it was necessary to try to enunciate a theory about the basic working of free choice. His solution to the dilemma of grace and free will is to accept—as had become firm orthodoxy—that grace is necessary and unmerited. Where he departs from Augustine is in denying that it is irresistible. Grace consists in God's restoring to the human agent the desire for justice which was lost. So grace is indeed the bestowal of values which the agent did not previously possess. The crucial choice facing the human agent to whom God has given the desire for justice is essentially the same choice that Anselm places before Satan.[26] The human being who has been given grace can hang onto it, or throw it away by willing some inappropriate benefit. (The AA example does not quite fit Anselm's actual schema for grace. To be analogous to Antti, AA had to hold the new values in an unsheddable way. Anselm reconciles grace with free choice by arguing that the agent can jettison his desire for justice.)[27] In

[24] Matthew 25:14–30; Mark 12:41–4; Luke 21:1–4; 1 Corinthians 12:4–30. Thomas Nagle in "Moral Luck" (1979) considers the fact that different people find themselves in different situations to be one aspect of responsibility-undermining moral luck. But in that any other state of human affairs seems quite impossible to imagine, it is perhaps a little odd to suggest that the lack of homogeneity in the human condition poses a reason to deny responsibility to human agents.

[25] Rogers (2008): 30–52.

[26] Note, though, that the fallen agent to whom grace is given is in a radically different situation from that of the unfallen agent. The individual and the whole human race have been infected by original sin. With the situation at the fall it is—metaphorically—as if the healthy person chooses to infect himself with a terrible disease. With grace—again metaphorically—it becomes possible that the agent not die of the disease. But nonetheless the agent remains seriously weakened due to the aftereffects of having been so ill. So when I say that Anselm holds that the choice for the agent who has received grace is essentially the same as for Satan before his fall, I am referring only to the restoration of the desire for justice, which allows the agent to confront the two morally significant options which Anselm takes to be required for an *a se* choice.

[27] Rogers (2008): 136–41. There is a superficial similarity here between Anselm's doctrine of grace and Mele's (1995: 172–3) assessment of agents who have suddenly received values that are new to them. Anselm agrees with Mele that the new values (in this case bestowed by grace) must be "sheddable" for the agent to be free. But, of course, for Anselm this is because options are required

Chapter One I noted that the critic of libertarianism might argue that accepting libertarianism has the unwholesome consequence of encouraging pride. My response was that on the Anselmian schema the free agent can take only a little credit for his choices. Anselm's doctrine of grace is a prime example. All the agent can take credit for is hanging on to what was given to him. But he *can* take credit, because the hanging on is entirely up to him.

The example of AA does not fit precisely with Anselm's doctrine of grace, but it is not too far off. AA, by the grace of God, has rejected his bad old views and now cares for life. But he is not (pace Augustine) home free. He still has morally significant open options. Perhaps, if he had chosen to hide out for the duration, that would have been a first, small step towards creating a bad character. He could no longer be a Nazi. His unsheddable new values close off that option. But he could be a coward. My point here is not to present a disquisition on grace, but to note that the Christian philosopher is used to the thought of sudden conversion. When St. Paul gets knocked off his ass on the road to Damascus, he is instantly changed from the avid persecutor of Christians to The Apostle to the Gentiles. Conversion stories are ubiquitous in the history of Christendom, and the Christian assumes that grace is at work—if usually less dramatically than in Paul's case—in the world today. And most Christian philosophers, whether libertarian or compatibilist, do not assume that grace destroys free will. The intuition that AA is indeed free may come more easily to the Christian, or someone who was raised Christian, or even just familiar with Christian history, thought, and culture, than to someone unfamiliar with, or hostile to, things Christian. Within the Christian milieu, sudden, value-reversing conversion, induced by a "manipulator," which conversion does *not* destroy freedom, has long been a standard part of the human story.

If those unconversant with Christianity share my intuition about the AA example, and are convinced of the plausibility of Anselmian internalism, so much the better. I do not see anything in the example that must prove a stumbling block to the naturalist, except for God. If we say that AA witnessed Fr. Kolbe's sacrifice and then fell and hit his head and awoke with the new values, and confronted the same choice, my intuition that he is praiseworthy for joining the Polish Underground does not change. I raise the point about the background

for aseity. One cannot be just unless one clings to justice *a se*. Mele allows that one may be causally determined either to retain or to shed the new values. The Anselmian holds that the retaining or shedding that Mele has in mind is not up to the agent in the right way, and, without aseity, would not ground or enhance the free will of the agent.

Christian beliefs as an example of the methodological proposal, mentioned in the introductory chapter, about noting how background beliefs may affect intuitions.

Degrees of Responsibility: Some Unpleasant Entailments

In defending Anselmian internalism I do not intend to dismiss the historical development which Mele takes to be crucial for autonomy. This issue was mentioned in the previous chapter in the section on character development, but there is more to be said here in connection with the internalist point that one's history is not the locus for the criteria required for freedom and responsibility. One thing the AA example shows is the appropriateness of ascribing degrees of responsibility. (If it seems better to say "degrees of praiseworthiness or blameworthiness," then understand "degrees of responsibility" to stand for the more unwieldy phrase.) So, for example, a young teenager—assuming he can make *a se* choices and so can bear at least basic responsibility—should be blamed for a crime he commits, but likely he should not be blamed as much as a normal adult should be. An agent who meets the Anselmian requirement for basic responsibility *and* Mele's pretty demanding criteria for autonomous agency should surely be thought to be more responsible for the actions that flow from the values he has embraced and the character he has built for himself over the years than AA, who is almost a child in terms of the new life he has embarked upon. The Anselmian is happy to admit that the sorts of capacities and practices which Mele's historical account insists upon are extremely valuable in enhancing the degree of responsibility of a basically responsible agent.

The other side of this coin is that the Anselmian must grant that an agent may be genuinely responsible, yet in a severely limited way. As the AA example illustrates, the Anselmian system entails the possibility that in some instances of responsible choice, we have little, if any, control over the factors which motivate us. Two unpleasant consequences of the position follow. (There may be more, but these two seem especially obvious.) The first is that it looks to be possible for an agent who has had only a little control over the formation of his character to nonetheless be basically responsible for his character-determined choices. One example of this possible scenario might be the agent who makes a few Anselmian choices as a child.[28] And these choices form his character to such

[28] Perhaps we make minimal self-forming choices—choices with some value content, open options, and a subsequent effect on our characters—much earlier, but surely, if we make them at all, most of us will have made some by our early teens.

an extent that he is precipitated along a "pre-set" path, freely and responsibly, but in a way that admits no future deviation, by choices he made when he first began to choose and had little, if any, control over the motivating factors. In that case he would be basically responsible for his character and his character-determined choices, but presumably only to a very limited degree.

As I noted in Chapter Four, just how and when our choices affect our characters seems an empirical, not an a priori, matter. One does hear of the rare agent, like Beth or AA, who changes his entire life plan due to some singular and striking experience. It seems likely, though, that most of us build our characters laboriously through years of making small choices, with a certain amount of tearing down some parts of the edifice and shoring up others. It is not clear that the Anselmian should absolutely rule out the possibility that an otherwise normal human being could live an apparently (at least from the third-person perspective) normal human life and *never* engage in *a se*, self-forming choices. (The Anselmian would not consider such a person to be a responsible agent.) The process of character construction is likely to vary significantly from life to life. So the Anselmian should allow the possibility of an adult agent whose choices are entirely produced by his character—he follows a fixed "character-path"—which character was entirely formed in childhood.[29] And it is a disquieting consequence that we may properly hold an adult agent responsible based ultimately upon choices he made as a child when he bore little, if any, responsibility for the options which were open to him to choose.

There are two points from which the Anselmian may take heart. First, we generally assume that only an agent who is rational at the time he chooses and acts can be responsible. But it is difficult to imagine an adult agent who is capable of rationally assessing his situation but who, because of his character formed in childhood, is consistently incapable of entertaining alternative, morally significant motivations. If agents *ever* do indeed confront the sorts of open options which the Anselmian envisions, it seems overwhelmingly likely that they confront them throughout their lives. An adult agent who is rational but who never struggles to pursue morally interesting, conflicting desires may be theoretically possible, but is so unlikely as not to pose any serious threat to the theory.

[29] Mele's externalist account, by insisting upon such robust historical conditions, rules out any remotely similar danger, though possibly at the cost of allowing very few into the circle of autonomous agents. Mele (2006: 129–32) does suggest that the libertarian might want to locate the scope of non-determined choices in childhood as a strategy for avoiding the luck problem. But it is not clear to me how this strategy—which I associate with the worrisome possibility under discussion here—helps with the luck problem.

But even the theoretical possibility of such an agent underscores the importance of the claim that responsible agents can be more or less responsible. How responsible you are for the construction of your character affects how responsible you may be for subsequent choices where the options, or even the choice itself, are determined by your character. Both the quantity and the quality of Anselmian choices an agent makes may be relevant to the degree of responsibility we may assign to an agent. In terms of character formation, an agent who has made numerous choices, some of which deal with very significant issues, is likely to be more responsible than an agent who has made only a few choices concerning less important issues. If that is plausible, we may be confronted with hard questions as we try to understand how to ascribe the appropriate degree of responsibility to an agent. How, for example, should we weight the importance of making numerous self-forming choices with little significance against fewer with more significance? My suspicion is that questions such as this are impossible to answer in the abstract. We would need to follow the course of a particular life with its various choices in order to assess how the interplay of externally produced motivations, self-forming choices, and character, affected the degree to which the agent can be held responsible. (This brings up a big practical problem, on which more below.) So it is an empirical question whether or not there are responsible agents who have succeeded in forming hard and fast characters at an early age. As I said, on an intuitive level it seems unlikely, even if theoretically possible.

A second point from which the Anselmian may take heart is that steps can be taken to avoid the possibility of the basically responsible adult agent whose character-path was fixed in childhood. As I noted, the Anselmian can happily agree that the sorts of capacities and practices Mele finds necessary for autonomy are, though not necessary for *basic* responsibility, at least important for enhancing responsibility. And you are just a better sort of agent—this, at least, is how the Anselmian sees it—if you are capable of bearing more responsibility.[30] So, for example, the extent to which an agent is self-reflective contributes to the extent of their responsibility and their value as an agent. A rational, adult agent who is truly self-reflective but who *always* follows a character-path fixed in childhood strikes me as not a live possibility. Of course, in addition to the metaphysical

[30] As I discussed in the introductory chapter, Anselm is writing within a theist worldview in which the point of human freedom is to become a better image of God. Mele writes as if being autonomous is a good thing, though I do not know that he would subscribe to anything like the sort of metaphysical value that Anselm assumes. It is not clear from Mele's discussion that, in the example where Chuck and Beth hold the same values, Mele believes the responsible Chuck to be in any way "better" than not-responsible Beth.

value of simply being a robustly responsible agent, there is also *moral* value to be considered. The wicked person is presumably less wicked if he suffers diminished responsibility than he would be if he were more responsible. Anselm would have it that a very wicked man is a better sort of thing than a very good dog, since the latter does not act responsibly at all. The best is to bear significant responsibility *and* to be virtuous. The best sort of life requires self-reflection and the other sorts of capabilities and practices that Mele endorses. The Anselmian can take heart, then, that parents and other educators can encourage the autonomy-producing habits of self-reflection etc. on the theory that the self-reflective agent cannot be propelled down a character-path chosen in childhood and can continue to grow in stature in terms of responsibility.[31]

In addition to the problem of the (merely theoretical?) possibility of the responsible agent whose character is determined by childhood choices, there is a second, related, unpleasant entailment of the Anselmian view. This one concerns justice, and is, alas, pressing in practice. (While practical problems do not show that a theory is false, when the question is responsible agency it is useful to appreciate how theory may be related to practice.)[32] The Anselmian, having allowed that sometimes the options confronting an agent are none of his doing, must grant that an agent may be more or less responsible based on the extent to which he has control over his motives and has responsibility for his character. This poses a daunting problem for the (non-divine) third party who—operating on the assumption that we do in fact bear at least basic responsibility—is tasked with assessing the extent to which an agent deserves praise or blame, especially as these translate into reward and punishment. The third party does not have access to the inner workings of the agent's mind, even when the agent stands before him, much less to the agent's life history as a self-forming agent. Even in the first person we may not have much insight. (Note that, while this problem may be especially thorny for Anselmian libertarianism, it is difficult to imagine any theory which includes responsibility-grounding free will and yet entirely avoids

[31] For a list of citations discussing psychological studies concerning how parenting styles can affect "self-organization" see William R. Miller and David J. Atencio, "Free Will as a Proportion of Variance." In *Are We Free? Psychology and Free Will* edited by John Baer, James C. Kaufman, and Roy F. Baumeister (Oxford: Oxford University Press, 2008): 275–95, at 283.

[32] As I noted in the introductory chapter, some philosophers couch the issue of freedom and responsibility in terms of determining the appropriate criteria for "us" to hold you responsible. Given that "we" suffer under such extreme epistemic limitations, the question of what it takes for you to actually *be* responsible is separate, though certainly related. In trying to spell out Anselmian libertarianism I am mainly addressing the latter question. And here, again, perhaps the difference in basic worldviews concerning who (or Who) might be monitoring your behavior is relevant to our intuitions.

the problem. Certainly externalist views like Mele's are subject to this practical difficulty as well, *mutatis mutandis*.)

To demonstrate the weight of the problem, take an example. Suppose an eighteen-year-old, Jules, is before the judge, and the charge of hitting an old man is proved against him. What punishment does Jules *deserve*? Suppose Jules has had a truly miserable upbringing. He was raised to be angry and violent. When the old man wandered into Jules' line of vision, his first thought was to slit the man's throat, just to take out his rage. But some glimmer of compassion from who-knows-where made him also want to attempt to control himself, if only a little. Say that, because of his motives, for which he does not bear any responsibility, Jules confronts two—and only these two—genuinely open options: either to hit the old man or to slit his throat. On the Anselmian account Jules is free and responsible. Say that Jules makes the better choice in hitting the old man. Perhaps he even makes his character just a little better than it was. Given all that we know—we who are privy to the facts about Jules' history and motivations—I take it that, oddly enough, we should say that Jules deserves praise rather than blame. It is not that he deserves praise *for hitting* the old man, but he deserves praise for choosing the better deed of the two options available to him.

Perhaps even odder, prima facie, is the point that a similar entailment holds for the (possibly imaginary) very good person who, on account of a sterling upbringing, is seldom if ever tempted to do wrong. We do not ere in calling that person "good" in some sense, but the Anselmian must say that he does not deserve a great deal of praise if it is not up to him that he is good.[33] He may be only a little bit praiseworthy if he chooses the better of two possible goods, and actually *blameworthy* if he chooses the lesser of two possible goods. To mitigate the oddness it seems right to suggest that this well-brought-up agent must make some self-forming choices in an effort to *conserve* and *build* the virtuous character on the foundation that was so well laid by his parents.[34] So perhaps the Anselmian should say that, in theory, someone who cannot help but be virtuous due to a good up-bringing might not deserve any credit, but, in practice, virtue always requires tending and deserves praise. The agent whose choices are limited to the good and the better may be able to take considerable credit for the fact that those are indeed his options.

But how much blame or praise does a given agent deserve? We know little about the minds and histories of those around us, so how can we ever judge

[33] This point constitutes the Anselmian version of the free will defense vis-à-vis the problem of evil.

[34] I thank Jeffrey Jordan for this point about conserving the virtuous character.

correctly? (Often we are blind even to our own condition.) Perhaps the best answer is that we can't. But this epistemic difficulty should not lead us to conclude that people are not in fact responsible, or even that they are likely to be less responsible than we usually take them to be. They may well be *more* responsible. Uninspiring though it seems, our best move, from an Anselmian perspective, may be to continue along the lines that we, as individuals and as a society, tend to follow already. We see that people should be held responsible, and when it comes to how harshly to blame them or extravagantly to praise them, we allow that what we can know of their histories should be admitted as evidence. Jules' lawyer should make every effort to demonstrate the ill effects of his unfortunate upbringing, and the thoughtful judge will take that into account. (Even if he could know *all* the facts, the judge, for consequentialist reasons, may have to sentence Jules to a punishment which he does not actually deserve.)[35] This process does not make for very exact earthly justice. The Anselmian is likely to be sad and cynical regarding even the more enlightened criminal justice systems. But unhappy and frightening consequences do not show that a view is false. If Anselmian libertarianism is a plausible theory, best to confront the problems generated by our limitations as agents and knowers and deal with them as we can.

And remember that, with all its difficulties, there is a practical advantage to embracing Anselmian internalism. If it is a good thing to be a responsible agent, then when we deny that our fellow human beings are capable of responsible agency we demean them. The Anselmian approach, though requiring the ability to make non-determined choices, otherwise sets the bar fairly low for inclusion in the community of responsible agents. AA makes the cut. Were Beth living in a non-determinist universe and, upon awakening with Nietzschean longings, were she to have the (non-compatibilist) option to resist and not commit the murder, then she would be responsible on the Anselmian account. People who are not grown up, and people who are not especially smart, may nevertheless be basically responsible. And good for them! Mele's compatibilist externalism, on the other hand, requires so much to be included in the history of autonomous agents that it seems likely that such agents will be few and far between. This is not evidence of the truth of the Anselmian approach, but it is reason to find it appealing.

[35] Katherin Rogers, "Retribution, Forgiveness, and the Character Creation Theory of Punishment." *Social Theory and Practice* 33 (2007b): 75–103, at 84.

6

Anselmian Alternatives and Frankfurt-style Counterexamples

> [Suppose] someone on his own (*sponte*) makes a vow to live in a holy community. . . . [H]e could be forced to keep it if he did not choose to: nevertheless, if he voluntarily keeps what he vowed, he is not less, but more, pleasing to God, than if he had not made the vow.
>
> *Cur deus homo* 2.5

> [Suppose Satan could choose only the good.] Therefore he would have a will that was neither just nor unjust. . . . [E]ven if he wills appropriately, it is not the case that his will is just on that account, since he received this [will] in such a way that he could not will otherwise.
>
> *De casu diaboli* 14

Introduction

In 1969 Harry Frankfurt proposed a telling counterexample to the claim that open options are necessary for the sort of freedom that can ground responsibility.[1] Frankfurt cites Mill as the originator (or at least an earlier proponent) of this sort of counterexample. Mill's example goes like this: Suppose someone, on his own, chooses to stay in a room. As it happens, and unknown to him, he cannot leave due to the door's being locked. Nonetheless he ought to be considered free. The fan of St. Anselm, however, is charmed to point out that, unless an earlier instance can be cited, the credit for originating the Frankfurt-style counterexample should probably go to Anselm, as the above quotation from *Cur deus homo* demonstrates. The monk, who can be forced to remain within the community—that is, he does not have the option to stay or leave—is nonetheless free and responsible if he voluntarily chooses to stay. Does this run counter to the claim, insisted upon in

[1] Harry Frankfurt, "Alternate Possibilities and Moral Responsibility." *Journal of Philosophy* 66 (1969): 829–39.

Chapter Three, that Anselmian libertarianism requires open options to ground the possibility for *a se* choice on the part of the created agent? Not at all. Anselm's position is clear, as Frankfurt's original statement of his counterexample is not, that the question of freedom and responsibility revolves around the nature of one's *choices*, not one's overt deeds.[2] What the example of the monk shows is that one can be free and responsible, so long as one has the ability to *choose* one way or another, even if one can be forcibly prevented from acting upon one's choices.

But Anselm—as the quote above from *De casu diaboli* shows—will not negotiate on the importance of options for *a se* choice. The created agent did not make himself, or produce within himself the original reasons and desires which motivate his choices, and so the only way the created agent could make *a se* choices is if he confronts genuinely open options such that it is entirely up to him which option he pursues. The Anselmian, then, must subscribe to a version of what is today often referred to as the principle of alternative possibilities (PAP). And pointing out that the issue is choices, rather than overt deeds, does not allow the Anselmian to evade the thrust of Frankfurt's argument. The principle behind Frankfurt's original counterexample has been restated to refer to PAP as it applies to choices, rather than overt deeds, and Frankfurt's challenge has evoked extensive debate in the last several decades. In that the requirement for open options is an ineradicable strand of Anselmian libertarianism, the Anselmian must respond to Frankfurt's attack. Happily, the Anselmian is in an especially strong position to do this. Indeed, I hope to show that the Anselmian version of PAP is simply immune to these Frankfurt-style counterexamples (FSCs). (I take it that my argument can be adapted by other species of libertarianism, *mutatis mutandis*.)[3]

The Anselmian argument against FSCs bears a family resemblance to a move made early in the discussion of FSCs, the "prior-sign dilemma defense" ("dilemma defense" for short). But the Anselmian approach offers a simpler

[2] A recent example of failure to make this important distinction is Roger Clarke's, "How to Manipulate an Incompatibilistically Free Agent." *American Philosophical Quarterly* 49 (2012): 139–49. The reader is told that an agent who cannot *do* other than a manipulator would have her do should not be judged responsible. However, in Clarke's example, the agent can *choose* other than the manipulator would have her choose. The reason she cannot *do* otherwise, is that the manipulator has set things up so that, when the agent chooses against what the manipulator wants, she does not actually engage in the overt deed, but only believes that she does. When she chooses what the manipulator wants her to choose, she both believes that she is engaged in, and actually engages in, the overt deed. The Anselmian has no doubt at all that the agent is responsible whichever way she chooses. Clarke's example is remotely similar to my "Rewind" suggestion at the end of the chapter, and will be discussed further in that context.

[3] In my *Anselm on Freedom* (2008: 73–81), I discuss FSCs in an Anselmian context briefly, but I had not then appreciated the impossibility of a Frankfurt-style scenario within an Anselmian analysis of freedom.

way to dismiss the FSCs. The dilemma defense assumes a limited controller, and its conclusions center on problems arising from that part of the Frankfurt story in which a limited controller is able to foresee that a certain choice is coming. This leaves the dilemma defense open to more recent responses involving "blockage" and "buffered" FSCs which aim to neutralize the problem of the controller's foreknowledge. But the discussion is advanced if we consider FSCs in an Anselmian spirit in which God plays the role of the hypothetical controller. (The basic argument could proceed without God, but casting it in theistic terms renders it more vivid.) God is omniscient and knows what you will choose tomorrow without needing any preceding evidence. Thus we can set aside concerns related to an epistemically limited controller and focus instead on a key thesis entailed by Anselm's insistence upon aseity—the grounding principle discussed in Chapter Four: The truth of a proposition about a free choice is absolutely dependent upon the agent's actually making the choice. This thesis seems to be assumed in the background of some of the various versions of the dilemma defense, but is not isolated and emphasized. Bringing it to the fore allows for a more direct criticism of the original family of FSCs than has appeared in the literature to date, and, with the addition of some plausible claims about what a free choice must entail, shows how PAP can be defended against the "blockage" and "buffered" FSCs.

After a very quick sketch of the FSC strategy and the dilemma defense, I briefly review some of Anselm's concerns and basic analysis of freedom, emphasizing points relevant to alternative possibilities and discussion of FSCs. I then explain how the Anselmian version of PAP is not susceptible to FSCs, even in their recently revamped iterations. I conclude that, on the Anselmian analysis, there looks to be one conceivable scenario in which a divine controller could engineer the situation so that an agent freely chooses the one option which the controller would have him choose. This scenario, however, does not involve FSCs, is metaphysically impossible, and probably cannot include genuine *a se* choice.

FSCs, the Dilemma Defense, and Responses

Harry Frankfurt attempts to undermine PAP through Frankfurt-style counter-examples. FSCs purport to show, in a way that connects with the concerns of the libertarian, that an agent who lacks alternatives may nevertheless be free and responsible. There is a long and winding literature on FSCs, but a quick sketch of the argument and the dilemma defense will suffice to make a start.[4] A typical (and

[4] For a volume devoted to the issue, see *Moral Responsibility and Alternative Possibilities* edited by David Widerker and Michael McKenna (Hants, England: Ashgate Publishing Limited, 2003).

simple) FSC might go like this: Suppose that a mad, but extremely knowledgeable and powerful, neurosurgeon (N) is monitoring Smith's brain. N intends for Smith to murder Jones. If Smith chooses on his own to murder Jones, and subsequently murders him, N does not have to do anything. But if N sees that Smith is going to choose not to murder Jones, he, N, will exercise control over Smith's brain causing Smith to choose to murder Jones. As it happens, Smith chooses, on his own, to murder Jones and murders him. N, then, simply stands by and does nothing. Given that in the actual sequence of events the controller doesn't do anything, the example is taken to show that Smith is free and responsible, although he could not choose other than to murder Jones, and so had no alternative.

The dilemma defense of PAP (versions of which have been proposed by Ginet, Kane, and Widerker) notes that, in the FSC, N must be able to foresee what Smith is going to choose. Presumably N must observe some sort of prior sign—a twitch or a blush perhaps—to provide the evidence for N to know what Smith is going to choose. The dilemma is this: If the prior sign provides incontrovertible evidence that the choice will occur, this can only be because the choice is causally necessitated by something prior to it. In that case the choice is not made freely. But if the prior sign is not followed inevitably by a given choice, then it is open to Smith to choose to murder Jones, or to fail to murder Jones, up until the time he chooses. In that case N does not have time to preempt Smith's choice if Smith should opt not to murder Jones.[5]

In response, defenders of FSCs, such as Mele and Robb, have proposed "blockage" variants which take N out of the picture. Since the epistemic limitations of a would-be controller are not at issue, the dilemma defense cannot gain any traction. As a rough example of a blockage argument suppose there is a "process"—P—in place and operating before Smith makes his choice. P aims at causing the choice to murder, unless Smith himself causes it. If Smith chooses not to kill Jones, then P, at that very moment, causes Smith to choose to kill Jones. But if Smith decides on his own to kill Jones, the process is preempted, and P does not produce any effect. Say that it happens that Smith chooses on his own to kill Jones, so that P does not cause Smith's choice. Isn't Smith free and responsible although he had no alternative but to choose to kill Jones?[6]

[5] Carl Ginet, "In Defense of the Principle of Alternative Possibilities: Why I Don't Find Frankfurt's Argument Convincing." *Philosophical Perspectives* 10 (1996): 403–17; Robert Kane, *The Significance of Free Will* (Oxford: Oxford University Press, 1996): 142–4; David Widerker, "Libertarianism and Frankfurt's Attack on the Principle of Alternative Possibilities." *Philosophical Review* 104 (1995): 247–61.

[6] Alfred Mele and David Robb, "Rescuing Frankfurt-Style Cases." *Philosophical Review* 107 (1998): 97–112.

David Hunt, defending FSCs, has proposed a "buffered" case which, while it retains the would-be controller, sets aside the problem of his foreseeing what Smith will choose. Say that N wants Smith to murder Jones at t2. And N knows that Smith will certainly decide to murder Jones *unless* Smith considers q at t1. Perhaps q involves moral principles or empathic feelings. For purposes of the example q just has to be something such that, if Smith were to consider it, it would make it uncertain whether or not he will choose to murder Jones. If N sees that Smith considers q at t1, he will step in and make him choose to murder Jones. Smith *could* consider q at t1, but as it happens he does not. N does not step in and Smith—having failed to consider q—chooses on his own to murder Jones. N, without any foreknowledge of Smith's choice, closes off the alternatives, yet should we not consider Smith free and responsible?[7]

The Anselmian argues that the FSC situation, in which the agent is free in the relevant sense, but has no alternatives, is simply impossible. It is impossible in its original formulation, impossible in its blockage and buffered versions, and impossible even if we replace the limited controller with an omniscient God. In fact, the impossibility can be shown most simply and directly when we hypothesize a would-be controller who knows, without any prior evidence, what it is you are going to choose.

Aseity and Alternative Possibilities

A quick review of Anselm's system and the concerns which motivate it will bring out key points relevant to the discussion of FSCs. Anselm hoped to show that created agents can be self-creating and they can have moral responsibility—genuine praiseworthiness and blameworthiness—in a universe made and sustained from moment to moment by an omnipotent, omniscient, all-good God. He held that there must be some lacuna in God's almost ubiquitous causation, to leave space for the created agent to act on his own, *a se*. If everything is God's doing, then there is no room for self-creation and responsible agency on the part of human beings. (I argued in Chapter One that the same problem for human responsibility arises if, instead of God, we hypothesize determinism and assign ubiquitous causal activity to a naturalistic universe.)

The value of moral responsibility explains why God permits wicked choices. It is good that created agents be capable of moral choice, but, Anselm argues, choice *must* entail options, including the live possibility that the agent choose badly.

[7] David Hunt, "Black the Libertarian." *Acta Analytica* 22 (2007): 3–15; "Moral Responsibility and Buffered Alternatives." *Midwest Studies in Philosophy* 29 (2005): 126–45.

God cannot engineer things to ensure that agents choose well without undermining their responsibility and their metaphysical stature.

To elaborate on the point noted at the beginning of this chapter, Anselm would apparently agree with Frankfurt's criticism of PAP if the principle were only about being able to perform alternative overt deeds. In fact, Anselm offers what may be the first stab at an FSC, explicitly aiming to undermine PAP understood to be about overt deeds. In his dialogue, *Cur deus homo*, he argues that God, being good, *must* become incarnate and die to save mankind. His interlocutor worries that God would then be unfree. Anselm notes certain unique features of divine freedom, but he also offers an example involving a human agent which is rather like an FSC. He aims to demonstrate that a lack of options, entailing the necessity of doing a particular act, does not conflict with the agent's freedom and responsibility vis-à-vis that act. He considers the monk who freely stays true to his vows and chooses on his own to remain within the monastery. Were the monk to try to leave, he would be physically detained, so he does not have the option to leave. (For this to be a true FSC, the monk should not be aware that he cannot leave, but Anselm's example does not achieve that level of detail.) But the monk is praiseworthy because he chooses on his own to stay.[8] So Anselm explicitly rejects PAP, if the principle is about overt *deeds*, and does so using a sort of proto-FSC. (With regards to the will of God, Anselm holds that He "must" choose and do the best.)[9] But when it comes to created *choices* Anselm insists that there must be options.

Anselm makes his case (spelled out in detail in Chapter Three) in an especially vivid way by asking how Satan, made good by God, came to choose to sin. He says that one theory is that Satan was subject to the weakness which being made from nothing entails, and so was motivated towards choosing badly. On this theory the decisive factor which explains why Satan sinned when other angels did not is that God did not give him the extra help he needed to hold fast to the good. But that can't be right, says Anselm. If Satan sins due to motivations which lead inexorably in the wrong direction, motivations which ultimately come from God as the creator, then the responsibility must be laid at the divine doorstep, and Satan cannot be to blame. Impossible![10] But where, in a universe made and sustained by God, is there room for the sort of freedom which could ground responsibility?

As Anselm sees it, the most basic requirement for the morally significant freedom which can ground moral responsibility is *not* alternative possibilities, but rather that one's choices come from oneself, they are *a se*. In modern parlance

[8] *Cur Deus Homo* 2.5. [9] Rogers (2008): 185–200. [10] *De casu diaboli* (DCD) 2.

we might say that Anselm proposes a version of "source incompatibilism." So, for example, God is free under the same definition of "freedom" as created agents, but He does not choose between alternative possibilities. He inevitably wills the best, but He wills freely in that His acts of will come from Himself in an absolute way because He exists absolutely independently.[11] God does not need open options in order to will with aseity. But created agents, including their faculty of will and their motivations, are caused by God. How can an agent which exists absolutely *per aliud*, through another, choose with aseity? If the agent is made to choose whatever he chooses by factors outside of himself—for example, motives of which he is not the author, or the immediate will of God acting upon him—then it was not really up to him what to choose, but up to these other factors. In Anselm's universe, if the created agent does not have aseity, then his choice is ultimately up to God, he is not an elevated *imago dei*, and it is just unfair to praise or blame him for what God made him do.

Anselm's thesis is that if the created agent confronts alternative possibilities such that it is absolutely up to the agent which alternative to choose, then the agent has a measure of aseity. So God endows the agent with two sorts of motivations, for justice and for benefit, which can come into conflict in morally interesting ways. The *a se* choice occurs when an agent is torn between conflicting motivations and ultimately opts for one over the other. This "opting" is absolutely up to the agent and constitutes the locus of aseity. As a different way of stating the same conclusion Anselm proposes a thesis which is radical in his day; the created agent is capable of making himself better than he was made by God. How is that possible in a universe in which all goods come from God? Anselm argues that, by choosing, on his own, to cling to the good that God has given, when he could throw it away, the created agent enhances his moral stature. Of course, in throwing away the good when he could cling to it, the agent sins. Alternative possibilities, then, are *absolutely necessary* if the point of creaturely freedom and responsibility is that the created agent should have the special status of being a creature who can, albeit in an extremely limited way, have some control over the direction of his life and the construction of his character.[12]

As developed in Chapter Four, Anselm does allow that choices which are determined by character may be responsibly made, on the assumption that earlier *a se* choices affect character such that a later choice determined by character is ultimately causally traceable to the sort of *a se* choice described above.[13] This means that not every responsible choice involves alternative possibilities.

[11] Rogers (2008): 185–200. [12] Rogers (2008): 73–8, 99–101.
[13] See also Rogers (2008): 83–5.

Nonetheless responsibility requires that at some point in his history the created agent engaged in *a se* choice and confronted alternatives. For present purposes it is this *a se* choice which is the focus and to which the term "choice" will refer.

Anselm, as discussed in Chapter Five, is an "internalist" regarding the criteria for moral responsibility. Whereas the "externalist" holds that it is the agent's history leading up to the choice that is the basis of responsibility, Anselm holds that it is the actual structure of the choice. In his universe, God (and possibly other free agents) supply everything concerning the agent's history up until the time the agent begins to make *a se* choices, and so there is no locus for aseity to be found in the agent's past. Anselm has to grant that—at least early in his career as a chooser—very little in the agent's choice is due to himself—just the opting for this over that.[14] But a little aseity is better than none, which is what there would be in a universe in which God (or natural processes, or malicious manipulators, etc.) caused everything about the agent including his choices.

Anselm recognizes that insistence on alternative possibilities entails the troubling consequence that the ultimate choice for this over that cannot be explained by appealing to anything in the agent's past or character. Indeed, while the agent can give reasons for his choice either way by pointing to the motives which were in play, there just is no sufficient explanation for the choice of this over that. Regarding Satan, Anselm writes, "Why then did he will? . . . Only because he willed. For this choice had no other cause by which it was by any means impelled or drawn, but it was its own efficient cause, and effect, if such a thing can be said."[15] But the conceptual discomfort of positing an event which cannot be fully explained by the past is the price to be paid if created agents, rather than God, are to be assigned the ultimate responsibility for their choices.

Anselmian Choice: Four Theses and a Timeline

Having reviewed Anselm's argument for the need for alternative possibilities as the ground for aseity for a created agent, we can refer to the *Anselmian* principle of alternative possibilities, APAP. APAP entails four key theses which need to be set out in order to see why it is immune to FSCs. Except for Thesis (c), these have been spelled out earlier in the book, but it is useful to have them collected and

[14] As discussed in Chapter Five, Alfred Mele (2006: 52) has criticized this approach, arguing that, since the agent himself is not responsible for the motivations in play, his ability to choose between them does not contribute anything worth having in the line of freedom and responsibility.

[15] DCD 27.

labeled here. The agent in question here is assumed to be rational and aware of his situation.

> Thesis (a) In order for an agent to make a morally responsible, *a se* choice the agent must first be motivated towards at least two, mutually exclusive, morally significant options. (From now on I will say "two" for simplicity's sake.) That is, he must be in TC, the torn condition described in Chapter Three.
>
> Thesis (b) The agent's choice between these options is absolutely up to the agent, which entails that it is not causally or externally, non-causally (ENC) necessitated in any way at all. (See the introductory chapter for ENC necessity.)
>
> Thesis (c) An agent who makes a choice for one of the mutually exclusive options cannot, simultaneously, make a choice for the other option.
>
> Thesis (d) It is the fact of the agent's choice which grounds the truth of propositions about the choice and the possibility of knowledge of the choice. This is the grounding principle set out in Chapter Four.

These theses need development, and they can be best explained by including a simple timeline for an *a se* choice. Given the description of Anselmian free choice in Chapter Three, the timeline is obvious. Nevertheless, in discussing FSCs, it is very important to pinpoint just when the relevant events occur, and so having the timeline clearly presented here will prove helpful.

> At t1 the agent (S) struggles to pursue two mutually exclusive, morally significant options.
>
> At t2 S chooses one over the other option.
>
> At t3 (in a typical case) S engages in the overt action that follows appropriately from the choice.

Thesis (a) holds that agents do not choose without motivations, so S must be motivated by at least two, morally significant, conflicting interests, wants, or desires, only one of which can be pursued. This is not to deny that all sorts of activities in addition to *a se* choices might have moral significance. Still, it is the choice between options which allows the satisfaction of the aseity requirement, on the stipulation that—Thesis (b)—it is absolutely up to S which of these options is ultimately chosen. The first required step in the timeline for an Anselmian free choice then, is this: At t1 S is in the torn condition. He struggles to pursue two options. He wants to achieve object A and he wants to achieve object B. For simplicity's sake suppose here that "object A" constitutes the deed of A-ing. So when S wants to achieve object A, he wants to engage in the deed of A-ing. And similarly for his wanting to achieve object B. We can abbreviate "to achieve object

A" as "A" and similarly with B. So we can describe the proposed situation this way: In accordance with Thesis (a), at t1 S wants A and S also wants B. *Ex hypothesi*, A and B are mutually exclusive, and known to be so to S. S must choose one or the other.[16] Absent TC, the created agent is not in a position to choose with aseity, and so cannot make a choice for which he is morally responsible.

Note that the alternatives are symbolized as A and B rather than as A and not-A. By hypothesis B entails not-A, but it presents a better picture of the Anselmian view of the process of making a free choice to oppose B, rather than not-A, to A. The Anselmian idea is that the agent is struggling to realize two, mutually exclusive desires. If I want to A—eat the chocolate cake, let's say—my struggle is not between wanting to eat the cake on the one hand and, on the other, wanting to not-eat the cake *simpliciter*. Rather, the Anselmian picture plausibly holds that the not-eating option is viable because I have a robust desire for some (real or perceived) good—losing weight, not spoiling my appetite, avoiding gluttony, etc. So, yes, the choice can be characterized as being between A and not-A; in the cake case I eat or I refrain. But it is important to recognize that the refraining is motivated by the desire for some good. We are not seriously torn in morally interesting ways unless we see mutually exclusive "goods" to pursue. Much of the free will debate is driven by intuition, and so it is helpful to include, in our picture of choice, salient features which might impact our intuitive responses.

The second step in the timeline of an Anselmian free choice is the choice itself. Say that at t2 S chooses B. T2 need not be instantaneous, though we often speak of the "moment" or the "point" of choice. It may be a temporally extended process. (In Chapters Three and Four I argued that the choice per se should not be thought of as a new "thing" with ontological status, but I do not believe that this entails that the choice be instantaneous.) The choice must, though, have a certain homogeneity. The act of choice for which the agent is to be held morally responsible cannot be supposed to contain earlier and later parts such that there is a process of choosing which can be stopped after it has begun but before it is completed. To see this, let us say that S's choosing B is constituted by the series of sub-events w–z. If—Situation 1—S's engaging in w makes it inevitable that he must (barring accidents) engage in w–z, then, for purposes of moral evaluation, he has made his choice for B by engaging in w, even if the whole choice is constituted by w–z. If the world ends after S engaged in w, even though S didn't get to x–z, God holds S responsible for choosing B. If—Situation 2—upon S's

[16] Can't S choose neither? In fact, in the Anselmian system, barring accidents, choices must follow motives, and in the case proposed S does not want to pursue a third option which is neither A nor B.

engaging in w, or w–x, or w–y, it is still open to S to refrain from completing all of w–z, and it requires w–z to constitute the choice, then S has not engaged in the entire process until S engages in z, and so it is not until S engages in z that S has chosen B, and it is open whether or not S will choose B until S engages in z. Choosing B may require w–z, but S has not chosen until S engages in z. Thus the actual choice, if not instantaneous, must nonetheless have the right sort of homogeneity. In Situation 1 the choice is homogenous in that x–z follow deterministically upon w. In Situation 2 the choice is homogenous in that it is not actually complete until the unified sub-event z. (This assessment stands, whether or not we suppose human choices to be physically constituted by the behavior of neurons in the brain.)[17]

Thesis (b) entails that nothing about the universe at any time (beyond the actual fact of S choosing B at t2) causally or ENC necessitates that S chooses B at t2. "The universe" here includes God. And "any time" includes the present and the future as well as the past. Philosophers often speak as if the factors which would undermine freedom must be *past* phenomena, but that is not the case. If God were, at t2, to cause S to opt for B at t2, that would not make any difference to the judgment that S is not to blame for a choice caused in him by God.[18] Or, if a mad neurosurgeon were, at t3, to send signals back in time to S's brain at t2, and cause S to choose B, the fact that the causal necessitation which produces S's choice at t2 originates in S's future does not affect his lack of aseity and responsibility. In order for S to choose B responsibly, it must be absolutely up to S that S chooses B at t2.

Thesis (c) entails that, in our example, S cannot simultaneously choose A and B. This "cannot" is very strong, maybe a logical "cannot." We assumed that our agent was rational and aware of his situation. Since B entails not-A, S should find it psychologically impossible to "choose" A and B simultaneously. Moreover, an *a se* choice consists in pursuing one option *rather than* the other. So it is logically impossible for S to be making an *a se* choice between A and B, and to be

[17] For an attempt to construct FSCs along the lines of the extended-event model see Eleonore Stump, "Alternative Possibilities and Moral Responsibility: The Flicker of Freedom." *The Journal of Ethics* 3 (1999): 299–324 and "Moral Responsibility without Alternative Possibilities." In Widerker and McKenna (2003): 139–58. In response to Stump see David Widerker, "Blameworthiness and Frankfurt's Argument against the Principle of Alternative Possibilities." In Widerker and McKenna (2003): 53–73, at 56–8.

[18] Hugh McCann has argued that God's immediately creating and sustaining an agent, including that agent's choices, is consistent with the agent's having libertarian freedom; "Sovereignty and Freedom: A Reply to Rowe." *Faith and Philosophy* 18 (2001): 110–16. McCann holds that the view he espouses is shared by Thomas Aquinas. I have criticized this position, most recently in "Anselm Against McCann on God and Sin: Further Discussion." *Faith and Philosophy* 28 (2011): 397–415.

"choosing" both A and B simultaneously. This point may seem too obvious to bother with, but it will turn out to be important in responding to FSC-derived challenges to APAP.

Steps one and two in the timeline are all that is needed for an *a se* choice, and, as I noted in the introductory chapter, it is in the choice where the moral action lies. But it is useful to include that in ordinary circumstances choices are likely to be followed by a third step, the carrying out of the deed.[19] Barring unforeseen occurrences, such as S getting hit by a truck or attacked by a mad neurosurgeon, we can suppose that at t3 S engages in the deed of B-ing. (T3 stands for some time after t2, but not necessarily the next moment. There might be a considerable lapse of time between the effective choice—that is the choice that leads to overt action—and the subsequent deed which follows from it.) But the morally relevant action has already taken place earlier, at t2. We may describe S's B-ing as morally responsible only if his choosing B meets the criteria for a morally responsible choice.

The Grounding Principle and Divine Foreknowledge

Continuing with our example of S choosing B at t2, Thesis (d), the grounding principle, entails that the truth of "At t2 S chooses B" depends on S's actually choosing B at t2, as does the possibility of anyone, including God, knowing that at t2 S chooses B.[20] Indeed, the grounding principle could be understood as just another way of stating the aseity requirement. The agent can be responsible only if it is absolutely up to the agent himself which choice he makes. So what is the case regarding the agent's choice is grounded wholly and only in what the agent actually chooses. This might seem too obvious to state, and, in proposing the dilemma defense, Ginet, Kane, and Widerker may well assume it. Kane, for example, writes that, on the assumption that a future choice is not determined, a would-be controller of free choice, "cannot tell until the moment of choice itself whether the agent is going to choose A or do otherwise."[21] Kane's point seems to presuppose the grounding principle or something very like it. But, to my knowledge, this point has not been isolated and underscored. It is important to state

[19] This schema is overly simplified. One important point ignored here is the fact that some of our most morally significant choices do not issue in deeds at all. Think of cases where one wrestles inwardly with vices, beliefs, emotions, or attitudes. For example, one may be torn between forgiving someone or nursing one's anger. Whichever way one chooses is likely to affect one's behavior eventually, but there may not be a direct causal line between this choice and some particular deed.

[20] For the historical Anselm on this point see Rogers (2008): 119–21.

[21] Kane cites his 1985 book *Free Will and Values* in "Responsibility, Indeterminism and Frankfurt-style Cases: A Reply to Mele and Robb." In Widerker and McKenna (2003): 91–105, at 91.

and emphasize it since it will play a major role in the argument for the immunity of APAP from FSCs.

A further reason for underlining the grounding principle is that, as noted in Chapter Four, there are those who are labeled libertarians, such as Ockamists and Molinists, who deny the grounding principle and suppose that there are true propositions about what a presently non-existent, perhaps never-to-exist, agent will freely choose. I concluded that the Anselmian's insistence upon aseity entails that a free agent's choice cannot ultimately be causally or ENC necessitated by anything outside himself, and so there cannot be true propositions about what he *will* or *would* choose in the absence of his actually choosing. How can there be divine foreknowledge, then? (The question is important here, since we will be invoking God's foreknowledge of free choice in demonstrating why APAP is immune to FSCs.) Anselm proposes the elegant solution of adopting an isotemporalist view of time. God knows what you choose tomorrow because tomorrow is immediately present to God, and He sees you *actually* choose what you choose. If S chooses B at t2, God knows this eternally, but the divine knowledge depends on S choosing B at t2 as explained in Chapter Four.

David Widerker has argued that divine foreknowledge conflicts with an agent having genuinely open options, and so it entails the sort of ENC necessity which is as inconsistent with free choice as is determinism.[22] Anselm grants that there is a sort of "consequent" necessity involved in divine foreknowledge. If God knows today that S chooses B tomorrow, then S cannot fail to choose B tomorrow. But in that God knows today that S chooses B tomorrow, because S chooses B tomorrow, the necessity originates with S's *a se* choice.[23] If we posit that S chooses B at t2, then it follows that it cannot fail to be the case that S chooses B at t2. It is the same necessity which accrues to knowledge of past or present choices. If God, or anyone, knows that you choose to read now, or that you chose to read yesterday, then you cannot fail to be choosing to read now, and you cannot fail to have chosen to read yesterday. In an isotemporal universe, what a given temporal perceiver calls the past, the present, and the future all have the same "fixity," because they all have the same ontological status. If this fixity involves a necessity which conflicts with free choice, then PAP would conflict with anyone ever freely choosing in the present or the past. And note that PAP and its variants are often stated in the past tense. We are often talking about

[22] David Widerker, "Blameworthiness and Frankfurt's Argument Against the Principle of Alternative Possibilities." In Widerker and McKenna (2003): 53–73, at 59–60.

[23] See Jeffrey Green and Katherin Rogers, "Time, Foreknowledge, and Alternative Possibilities." *Religious Studies* 48 (2012): 151–64.

praising or blaming someone for what they *have done*, and we say that it must be that the agent "could have done otherwise." But the past is fixed. If S chose B yesterday, it is today impossible that S failed to choose B. But surely this impossibility does not entail that S could not have chosen freely yesterday. The point is even more obvious with present choices. If S chooses B *right now* it is impossible that S not choose B right now. It is just absurd to argue that the necessity which is ascribed to the occurrence of a choice, just because the choice occurs, is the sort of necessity which could conflict with free choice. By adopting isotemporalism Anselm and the Anselmian can accept divine foreknowledge without holding that future choices must be causally or ENC necessitated.

On isotemporalism God knows what happens at all times because all times are "present" to Him. The agent can "choose otherwise" because he is torn between options, the choice is not causally or ENC necessitated, and it is absolutely up to him which he chooses. God can foreknow what the agent will choose only because He "sees" him choose it. On the Anselmian account, the "consequent" necessity which accrues to the choice beforehand, simply because it is foreknown, is completely innocuous vis-à-vis the agent's freedom and responsibility.

David Hunt, defending FSCs, writes that introducing God "is at least as effective as a perfect passive alternative-eliminator [like P in the blockage case mentioned earlier] in rendering the entire actual sequence unavoidable, leaving no alternatives that could support [the agent's] free agency under any reasonable version of PAP."[24] But on the Anselmian account God's knowledge does not *do* anything at all to "render the entire actual sequence unavoidable." Agents freely produce what happens and God knows about it. If Anselm's APAP is reasonable, then, allowing isotemporalism, it maintains robust alternatives in the face of divine foreknowledge.

Anselmian Immunity to FSCs

Can Anselm's APAP be defeated through some analog of an FSC, an Anselmian-relevant FSC, or AFSC for short? Let us try to construct one. In Anselm's universe we do not have to invent an impossibly knowledgable and powerful neurosurgeon to play our would-be manipulator, and we do not need to address the question of how this epistemically limited manipulator can foreknow what the agent will choose. An omniscient God is ready to hand. The AFSC, then, should go something like this: Say that God wants S to choose A at t2. If He sees that S is

[24] David Hunt, "Freedom, Foreknowledge, and Frankfurt." In Widerker and McKenna (2003): 157–83, at 174 ff.

going to choose A on his own at t2, then He will not take any action. If He sees that S is going to choose B, then He will step in and cause S to choose A. As it happens S chooses A on his own. God does not step in. Isn't S's choice of A free, although he could not have chosen otherwise?

But this AFSC is an impossible scenario. Logically impossible. Given Thesis (d), the grounding principle, if God sees that S is—in S's future—going to choose B at t2, this can only be because S chooses B at t2. Even God cannot do the logically impossible. So, if S chooses B at t2, even God cannot make it be the case that S does not choose B at t2, but chooses A instead. And, by Thesis (c), even God cannot make it the case that, if S chooses B at t2, S chooses both A *and* B at t2. God cannot leave S free and see to it that S chooses A at t2. APAP cannot be threatened by AFSCs. Any theory embracing Thesis (d) can deploy the same line of argument against any FSC that posits a counterfactual intervener who will step in if he foresees that the agent is freely *going to choose* other than the intervener would have the agent choose.

A similar argument can be made for any FSC that posits an intervener who will step in because he knows what the agent *would* freely choose in some not-to-be-actualized situation. Even God cannot know what S will choose, in the absence of S's choosing, and even God cannot know what S "would" choose in a non-actual situation, since there is no truth to propositions about non-actual free choices. This argument is similar to the dilemma defense of PAP against FSCs offered by Ginet, Kane, and Widerker, but it makes the case clearer by hypothesizing an omniscient would-be controller, and by setting out Thesis (d) as a key premise, rather than tacitly assuming it.

Mele and Robb have responded to the dilemma defense with a "blockage" scenario which neutralizes problems concerning the prior knowledge of the limited intervener. In place of the mad neurosurgeon who observes the agent and must foresee what his choice will be, they posit a process, P. In spelling out their scenario, for simplicity's sake, I replace their examples of the agent (Bob) with my S, and the desired decision (to steal Ann's car) with my A. They write, "The process, which is screened off from [S's] consciousness, will deterministically culminate in [S's] deciding at t2 to [A] unless he decides on his own at t2 to [A] or is incapable at t2 of making a decision... The process is in no way sensitive to any 'sign' of what [S] will decide. As it happens, at t2 [S] decides on his own to [A]... But if he had not just then decided on his own to [A], P would have deterministically issued, at t2, in his deciding to [A]."[25]

[25] Mele and Robb (1998): 102.

But the Anselmian responds that no mere process can succeed in doing what was impossible for God to do. The scenario, as Mele and Robb set it out, includes the claim that if S chooses other than A—that is, S chooses B in the Anselmian example[26]—at t2, then P makes it the case that S chooses A at t2. But if S chooses B at t2, it is impossible that he choose A, rather than B, at t2. And it is impossible that he choose A *and* B at t2. Mele and Robb attempt to flesh out the situation they envision by proposing a "neuro-fictional" picture of the two processes at work, [S's] indeterministic deliberative process, x, and the process that ensures that [S] chooses [A], that is, P. The picture involves two different "decision nodes" in [S's] brain such that, "The 'lighting up' of node N1 represents his deciding to [A], and the 'lighting up' of node N2 represents his deciding *not* to [A]."[27] The situation in which S chooses B on his own at t2 and P causes him to choose A at t2, then, is pictured as, "if, at t2, P were to hit N1 and x were to hit N2, P would prevail... P would light up N1 and the indeterministic process would not light up N2."[28]

But, in order to be the physical instantiation of an FSC which can connect with libertarian concerns, these neuronal goings on must map onto the conscious choices of the agent in the right way. Does the "hitting" of the node occur before and cause the "lighting up" which constitutes the choice? If so the "hitting" itself does not occur at the moment of choice and does not constitute the choice, but rather *produces* the choice for A or B. In that case the picture described is deterministic all the way around, with or without P in place, since the "hitting" is a neuronal event which is not the choice, which occurs before the choice, and which produces the choice for A or for B.[29] On the Anselmian account, if S chooses A freely, there is nothing before the choice which causes S to choose A. If it is the lighting up of N1 that constitutes S's choice of A, it cannot be caused

[26] Perhaps Mele and Robb could say that in their example the would-be options under deliberation—and one of which P ultimately closes off—do not consist in S's choosing to A or choosing to B, but rather they consist in his choosing to A or not choosing to A, or else in choosing to A or choosing to not-A, and somehow this shields their example from my argument. I do not see that it does, but in any case, if their example is significantly different from mine where S is deliberating between A-ing and B-ing, in that their agent is not debating between at least two options, then the Anselmian holds that it could not be an instance of a morally significant, robustly free choice at all.

[27] Mele and Robb (1998): 104.

[28] Mele and Robb (1998): 104. Here Mele and Robb try to make the neuro-fictional picture more plausible by drawing an analogy with a widget-making machine where the color of the widgets to be produced is triggered by bbs striking in cups. They develop the analogy further in "Bbs, Magnets and Seesaws: The Metaphysics of Frankfurt-style Cases." In Widerker and McKenna (2003): 127–38.

[29] In that Mele and Robb (2003) distinguish between the bb "striking the cup" and the "commitment" on the part of the machine to produce a certain color widget they seem to suggest that the hitting is prior to and causes the lighting up.

by a preceding, non-chosen event, and still be free.[30] But if the hitting and the lighting up are simultaneous, and constitute the choice, then it is impossible that N1 and N2 are hit simultaneously, unless Mele and Robb dispute Thesis (c) and hold that a rational and aware agent can knowingly opt for A over B and for B over A at the same time. They do seem to *say* that P's causing S to choose A at t2 is triggered by its not being the case that S chooses A at t2. And they insist upon these two events happening simultaneously. "Why can't *P* cause the decision *at t2*, the very time that the decision fails to be caused by *x*?" they ask rhetorically.[31] But, on anything resembling a robust, libertarian account of choice (assuming the principle of non-contradiction) it is difficult to see how this is possible. S's choice for B cannot be "blocked" at the same time that it is actually made.

Robert Kane offers a somewhat different response to the Mele/Robb attempt at an FSC which avoids the dilemma defense. He notes that, by closing off options—either all options, or all "robust" options—P essentially determines [S's] choice for [A].[32] The Anselmian agrees. If God engineers a blockage situation such that S can "choose" only A, then, even if S "chooses" A as a result of his own deliberations, this does not constitute a free choice. S's choice for A is *a se*, precisely because he is choosing A when he could have chosen B. The Anselmian could put the point this way: the supposed virtue of the FSC is that it includes the claim that the agent, on his own, chooses what the manipulator would have him choose, and so it leaves the "actual sequence" of the choice untouched. The intuitive power of the FSCs is derived from the claim that, since the manipulator did not have to step in, the "actual sequence" is the same as it would have been in a simple case of libertarian free choice with no manipulator in the wings. But this is false in the Mele/Robb example vis-à-vis an Anselmian choice (and probably more generally). The *a se* choice at t2 must be preceded by the agent's being in the torn condition (TC) at t1, struggling to pursue each of two incompatible motivations, either of which he could in fact pursue. The agent believes he is able to choose either A or B. Call this belief "q." In the case of an Anselmian choice, at t1

[30] As noted in Chapter Three, the Anselmian would dispute Mele and Robb's description of the libertarian free choice as one in which an indeterministic process *produces* a choice. Perhaps Mele and Robb's example gains more traction against Kane's event-causal account, where one of the preceding "efforts" probabilisitically causes the choice. But for the Anselmian it is the choice itself that is not determined. The preceding process might be determined up to the point of choosing, as in Anselm's paradigm case of Satan's original choice, where God has supplied the competing motivations which generate the struggle to choose.

[31] Mele and Robb (2003): 133; Mele (2006): 88.

[32] Robert Kane, "Responsibility, Indeterminism and Frankfurt-style Cases: A Reply to Mele and Robb." In Widerker and McKenna (2003): 91–105.

S is in TC, believes q, and q is true. In a blockage AFSC, at t1 S is in TC, believes q, and q is false. But it is one thing to have a true belief and another to have a false belief, so S is not in the same situation at t1 in the blockage example as in the original example, and the "actual sequence" cannot be the same.

Moreover, recall Anselm's basic motivation—to explain how a created agent, in a universe where all goods come from God, might be able to make himself better and be deserving of praise. He could do so, Anselm says, if he could cling to the God-given good *on his own*. But the only way that is possible is if he chooses to cling when he could abandon the good. He gets the credit for clinging to the good precisely because he could let it go. I remarked in Chapter Four that there is no *potiusitas*, no "rather-than-ness" as a property with ontological status attaching to a choice. And yet it is the "rather-than-ness" that allows the choice to be attributed to the created agent, rather than to God. Say that A involves choosing rightly. If God engineers the situation such that, at t2, S cannot choose B, then S does not choose A *on his own*, although phenomenologically it might feel to him as if he does. Thus he does not make himself better and is not praiseworthy. With the option for B blocked, the two moments of the "choice" in the blockage AFSC are quite different from what they were in the original example, and the "actual sequence" in the blockage AFSC simply does not retain the elements that made the original an *a se* choice.[33]

A somewhat different attempt to get around the dilemma defense is suggested by David Hunt who proposes a "buffered" FSC.[34] Adapted to the Anselmian case, Hunt's proposal goes roughly like this: God wants S to choose A at t2. S will choose A at t2 unless he considers q at t1. (Perhaps q involves reasons or beliefs that would render B a viable option.) If God sees that S considers q at t1, He will step in and make him choose A. S *could* consider q at t1, but he does not. God does not step in, and S chooses A on his own. S has no alternative in that, having failed to consider q at t1, it is inevitable that he choose A at t2. Is this a case where S is free and responsible without alternatives?

The Anselmian responds that this buffered scenario does not meet the criteria for an *a se* choice. S's "choice" for A in the buffered AFSC was not preceded by the torn condition. Here S wants only A and goes for it. The agent must actively recognize mutually exclusive options in order to make an *a se* choice. If, for reasons for which the agent is not responsible, the agent sees only one desirable object, then his pursuit of it does not constitute an *a se* choice. (Though

[33] This is not the same as the Reidian "agent-causation defense" in that Anselm's understanding of agent-causation is somewhat different from Reid's. See Michael McKenna and David Widerker, "Introduction." In Widerker and McKenna (2003): 1–16, at 8–9.

[34] David Hunt (2007) and (2005).

remember that the Anselmian allows that the agent may be responsible for character-determined choices if the character was formed through the sort of *a se* choices under consideration here.) If we assume from the outset that an agent need not actually consider alternatives in order to "choose" responsibly, then we may be satisfied that the buffered case presents an agent who can be responsible without alternatives. But for the Anselmian, who sees robust alternatives as requisite for aseity, this seems to beg the question.[35]

If the options are closed off *before* the choice, the agent does not make an *a se* choice. And it is just impossible that the options be closed off *at the moment* of choice. The conclusion is that FSCs cannot undermine the requirement for alternatives in Anselm's version of libertarianism. Is there any way at all that a would-be controller could insure a particular choice on the part of an agent without destroying the agent's freedom?

Rewind

Prima facie there might look to be one conceivable scenario in which it is possible for God to see to it that S chooses A at t2, and yet S's choice of A at t2 is free and responsible. Call the scenario "Rewind." God wants S to choose A at t2. If S chooses B at t2, God "rewinds the tape," restoring the whole universe (*sans* Himself) to its condition at t1, and waits (logically, if not temporally) to see what S chooses on this next iteration of the choice.[36] That seems logically possible. If S chooses A at this next iteration of t2, well and good. If not, things get rewound again. It is *logically* possible that S choose B at every rewind, even assuming an infinite number of rewinds. In order for this to be a case where God can be sure to get the result He wants Rewind must assume Aristotle's understanding of ontological possibility. Aristotle, and many subsequent philosophers, invoke the principle of plenitude, and hold that, given an infinite time, in Aristotle's words, "what may be, must be."[37] If it is (on this Aristotelian understanding)

[35] Hunt raises something like this criticism against himself (2005: 134–7), but his responses are unsuccessful against the Anselmian PAP, in that they tend to stipulate that agents not confronting recognized options can be responsible—the very question at issue. Pereboom proposes a somewhat similar FSC in which the issue is whether or not the agent achieves a certain level of attentiveness and cites Kane's response, with which I essentially agree; "Hard Incompatibilism and Its Rivals." *Philosophical Studies* 144 (2009) 21–33, at 28–9.

[36] Peter van Inwagen suggests the "rewinding" move as an intuition pump for the thought that indeterminism means luck or chance; "Free Will Remains a Mystery." In *Philosophical Perspectives 14: Action and Freedom* edited by James E. Tomberline (Oxford: Blackwell Publishing, 2000): 1–19.

[37] *Physics* 3.4 (203b30). J. Hintikka argues that Aristotle really means it; "Aristotelian Infinity." *Philosophical Review* 75 (1966): 197–218. It is this understanding of possibility that drives Thomas Aquinas' Third Way.

possible that S choose A, then, given enough iterations, S cannot fail to choose A. (Presumably God knows eternally how many iterations it takes, since He "sees" S actually choosing A on the *n*th iteration. But we are going to set that thought aside because it introduces a tangle which I am unwilling to confront. Assume we have a temporal God in a presentist universe Who must wait to see what S chooses.) No matter how S chooses originally, God eventually gets the universe in which S chooses A at t2. If S chooses A at t2 the first time around, isn't he free and responsible? Indeed, at least prima facie, his choice seems to meet the criteria for an *a se* choice made at any iteration of t2 (though more on this below).

In addition to the point that Rewind must assume an Aristotelian notion of possibility, which some will find uncongenial, there is a further problem with the scenario. The claim was that God gets the universe in which S chooses A at t2, no matter what S chooses originally. By hypothesis Rewind involves the possibility of at least a second iteration of the choice, which means a second iteration of S's being in the torn condition (TC), struggling to choose between two options, so there must be a second t1 which can be followed by the choice at a second t2. The second iteration is not numerically identical to the first. (S is not exactly the same either, in that S at the second iteration has the first iteration in his history. But presumably God, in His omnipotence, could restore S to his "original" condition in all relevant respects—"relevant," as discussed below, means in almost every way—so this does not undermine the possibility of Rewind.) There is a "new" t1 followed by a "new" t2. It is simply impossible, on an Anselmian account, for God to see to it that S, freely and responsibly, chooses A at t2 *simpliciter*. But in Rewind God has *almost* total control over the outcome. Call the iterations of t1 and t2, "t1* and t2*, t1** and t2**, t1*** and t2***"...etc. So even if God cannot secure the outcome at t2 (no star), given an infinite number of t-stars to work with, He can come close to securing the outcome He desires. God can see to it that S chooses A at t2, or t2*, or t2** etc. And so this is a case where the ultimate way the universe goes regarding S's choice for A is up to God, but if S chooses A at t2 (or t2*, or t2**, or t2*** etc.) he (apparently) chooses freely and responsibly.

Rewind is not a variety of AFSC in that it does not include God cutting off an alternative possibility. S cannot follow through with *doing* other than A, since the universe gets rewound if he chooses B, but, as I have noted, it is the ability to make an alternative *choice*, not the ability to do an alternative *deed*, that is required for aseity and responsibility.[38] But it is like an AFSC in that it constitutes a criticism of APAP by (apparently) showing that an agent can be free and

[38] See endnote 2.

responsible even if a manipulator can act to secure a desired choice. That seems to diminish the importance of the alternatives vis-à-vis the agent. And Rewind poses a challenge to Anselm's free will defense against the problem of evil. If God could maintain human free choice and self-creation, and yet secure whatever choices He deems best, why is there so much wickedness?

But Rewind does not actually undermine Anselm's theory of freedom or pose a serious threat to the free will defense. Although Rewind is conceivable, it is (at least in Anselm's universe) impossible, and there is a serious question about whether or not the situation in Rewind really maintains free choice. Note, first, that a lesser manipulator than God, or at least a very god-like being, could not accomplish the needful, since what is required is the ability to recreate the universe to produce the iterations of t1. To see this consider the scenario proposed by Roger Clarke.[39] In Clarke's scenario, there is a manipulator who would have an agent buy the (morally) bad eggs rather than the (morally) good eggs. The agent believes that the buying is accomplished by the agent's pushing a button, having made the choice regarding which eggs to buy. But when the agent does not choose as the limited manipulator wants her to choose, she pushes the button to buy the good eggs, but the button does not actually initiate the egg-buying process. In that case her memory is erased by a limited controller, and the agent again finds herself confronted by the egg-buying choice and the buttons. Eventually, one supposes, she chooses the bad eggs, pushes the button, and that button initiates the egg-buying process. So the agent cannot do otherwise than buy the bad eggs.

But (in addition to the fact that the Anselmian is focused on the *choice* not the overt deed) Clarke's scenario is ultimately quite different from the proposed Anselmian Rewind. Clarke's agent must suffer significant change, without the advantages of an omnipotent controller to restore her to her status quo ante. To be an Anselmian Rewind scenario the torn condition (TC) in any subsequent iteration must be the same in all respects as in the original situation, save for the fact that it is a subsequent TC iteration, and not the original TC. The Anselmian assumes that very small differences can affect the ruminations of the free agent. Do your feet hurt just a little more than they did a minute ago? Has the noise from the air conditioner begun to impinge upon your consciousness slightly? Etc. etc. etc. If the agent is in a somewhat different TC, then the choice is a somewhat different choice. The relevantly "same" choice requires that the agent in a subsequent iteration be the same as he was in the original situation in all respects, save that now he is in the iteration and not the original situation. And the

[39] Clarke (2012): 140–1.

situation in the subsequent iterations must be the same in all respects, save that now it is the iteration and not the original situation. Clarke's example does not satisfy these criteria. An agent whose memory has been erased, through some (theoretically) humanly possible, non-divine process, is very different from an agent whose memory is intact. An agent whose memory has been erased *twice*, is very different from an agent whose memory has been erased only once. An agent who is a few minutes older is different from an agent who is a few minutes younger. An agent who has been sitting (thinking, digesting, etc. etc.) for an hour is relevantly different from an agent who has been sitting etc. for an hour and a half. Our physical surroundings—which may be relevant to TC on the assumption that the brain is a very complex system, sensitive to slight changes—are inevitably different than they were a short time ago... the air pressure and temperature have changed, the content of the air has changed, the position of the moon has changed, as has the position of the earth, etc. etc. etc.

Clarke says that his agent is faced with "exactly the same choice again." However, due to the limitations under which his non-divine manipulator must operate, this is not true. (To make the point he is making, he probably does not need the agent to be in exactly the same TC. A series of somewhat different choices to buy the good eggs, which ends when the agent chooses to do what the manipulator would have her do, buy the bad eggs, would suffice for his example. This is because he is interested in shutting off alternatives regarding overt deeds, not regarding choices. But for the Anselmian, as bears repeating, it is in the choice that the moral action lies.) In order for a manipulator to succeed in having an agent make almost exactly the choice the manipulator wants him to make—"almost" to account for the fact that the choice in the rewound universe is not numerically identical to the choice in the original universe—the manipulator must be God or at least a being with god-like powers. The Anselmian holds that the choice must be preceded by an all but numerically—and perhaps temporally—identical TC. Otherwise it's just a different choice altogether. But a limited rewinder could not replace the universe to a situation identical to t1.

It is safe to say that *Anselm's* God could not participate in Rewind.[40] Rewind is contrary to the omnibenevolence of God in that it involves the massive deception of created agents in several different ways. If S chooses B at t2, God "revokes" the whole universe as it has come to be at t2, in order to rewind it to t1*. A first area

[40] We could not simply plug God in to Clarke's example, either, for reasons that are analogous to the reasons God cannot participate in Rewind. Could we imagine an evil god in charge of some universe and engaging in Rewind? No. An evil god is intrinsically incoherent—evil being a lack, or failing, or absence, or corruption of the good. Some being less than God cannot bring things into being, and that is essentially what is required for Rewind.

of ubiquitous, systematic deception would occur in that we all suppose that "time marches on," one moment following another, such that the universe is not being "reset" to a new iteration of an earlier time. So Rewind involves massive deception concerning how the universe goes on. Furthermore, S, and indeed all the citizens of t2 who are capable of having memories of t2, must forget t2 when the universe is rewound to produce t1*. Otherwise t1* would not properly replicate t1. And that widespread amnesia is deception. Third, and perhaps most importantly in context, in order for the choices in question to have the moral significance they are supposed to bear, created agents must not know about Rewind, even in advance. An agent who knew about Rewind would reasonably say to himself, "If God can see to the outcome in any case, it doesn't really matter what I choose." So Rewind involves at least these three types of massive deception. But deception is bad. At least it is in Anselm's universe where God is Truth with a capital "T," the source of all that is true and good.[41] God, if there is a God, is obviously able to permit deception in that people are often deceived. But in Rewind, since He is the rewinder, He would have to be causing the deception by rewinding the universe, erasing the memories of anyone or anything with memories, and generally deceiving the agents in question. That is impossible.[42]

And it is not clear that the iterated choices in Rewind have the robust aseity the Anselmian requires. An ENC necessitated choice is not *a se*. ENC necessitation exists when, even if a choice is not *causally* determined, the outcome of a choice is rendered necessary by something *not* the agent's actual choice. In Rewind there seem to be three elements, none of which are S's choice, or dependent on S's choice, which together render S's choice for A necessary at some iteration of t2. They are the fact that S's choosing A is possible, the Aristotelian version of possibility in which, allowing an infinite time, all possibilities are realized, and God's plan to repeat the universe until S chooses A. Thus it seems that, when S eventually chooses A, the choice is ENC necessitated and hence not *a se*.[43]

[41] See Anselm's *De Veritate*.

[42] Rather than Rewind, one could propose "Copier" in which God creates an infinite number of identical universes, all at t1. Assuming Aristotle's understanding of possibility, at some of the universes, S chooses A at t2 and at some S chooses B. God can then just destroy the universes at which S chooses B (and spare all the universes in which S chooses A?) and achieve the divinely desired result from a free choice. This alternative scenario suffers from the devastating difficulty that God must destroy at least one universe. For Anselm's perfectly good God that is an impossible suggestion. More on the similar suggestion of "branching universes" in Chapter Seven.

[43] In a Rewind scenario involving a non-Aristotelian understanding of possibility, it might well be the case that S never chooses A. In that scenario S's choice—whether for A or for B—at any iteration, assuming the Anselmian criteria are in place, should probably be considered *a se*.

But here is a question: Suppose the three ENC necessitating elements are in place before the original instance of t2. What if S chooses A on the first go round? There is no rewinding. Then it seems peculiar to say that his choice was not *a se*. Everything about S and the situation was exactly the same as it would have been *sans* Rewind. Since there is no rewinding if S chooses A in the original iteration, it seems wrong to say that the choice is not free. I think the most plausible thing to say here is that, if Rewind is never *put into effect*, then the three elements do not ENC necessitate the choice in the original iteration. The divine plan and the nature of (Aristotelian) possibility have no connection with the choice. If S chooses A originally, then it is everything an Anselmian free choice ought to be. But if God rewinds to get the choice He wants, then the choice for A is ultimately not free, since it is ENC necessitated. I am not sure what to say about subsequent choices for B. I tend to think they are free since they are not the outcome of the three ENC necessitating factors. But this is an odd situation. The choice for B is made responsibly, but not the (rewound) choice for A? Perhaps this is yet another reason to think that Rewind is ultimately just incoherent.

A final, and perhaps most telling, reason for thinking that Rewind is impossible in an *Anselmian* universe is that free choice is deeply important. Allowing the created agent a measure of aseity bestows the elevated metaphysical status of being an image of the divine. God made people free for a purpose and He does not abandon His projects. Rewind, even if my point above is mistaken and it does not render the rewound choices ENC necessary, nevertheless dilutes the importance of free choice. Rewind entails that agents—without their knowledge—are confronted with innumerable "do-overs" until they choose what God would have them choose. This undermines aseity and devalues freedom. Anselm's God would not participate!

On the Anselmian understanding, alternative choices are required for freedom and responsibility. AFSCs cannot undermine this claim. Rewind may be conceivable, but it need not concern the Anselmian since it is not possible in an Anselmian universe. For those sympathetic to Anselm's analysis of freedom and responsibility robust alternatives are here to stay.[44]

[44] A version of this chapter was read at the Big Questions of Free Will conference at St. Thomas University in St. Paul, Minnesota, September 2013. I thank participants in the conference for helpful comments, which led to significant revisions.

7

The Luck Problem

Part I. Probabilities and Possible Worlds

> Why then did he [Satan] will? . . . Only because he willed. For this willing had no other cause by which it was by any means impelled or drawn, but it was its own efficient cause, and effect, if such a thing can be said.
>
> *De casu diaboli* 27

Introduction

According to Anselmian libertarianism a created agent's ability to "self-create" requires that he be able to make *a se* choices, and that means he confronts radically open options. How, then, is an agent responsible for an *a se* choice, when there is nothing about him which explains why he chose one way over the other? The criticism is not that a libertarian choice is uncaused. To take our standard example of S choosing B, Anselmian libertarianism insists that, in terms of the motive power behind S's choice, it was the desire for B that caused the choice. And in terms of S's *preferring* B *over* A, well that was absolutely up to S, so S causes it by per-willing B as explained in Chapter Three. Most of the participants in the present debate, even those who oppose libertarianism, have allowed that being caused does not require being causally *necessitated*. So the issue is not that there is no cause at all for S's choice.

Nevertheless the Anselmian has to grant that, on the question of why S per-willed B when he really could have per-willed A, there is nothing more to be said by way of explanation than that he chose . . . by per-willing. And this is an uncomfortable point, as Anselm himself recognizes. In his dialogue on the fall of the devil his student presses him regarding why Satan fell while other angels who were, in all relevant respects, in *exactly* the same condition, held fast to the good. He responds with the quote at the top of this chapter; there is no cause beyond S's choosing, "if such a thing can be said." That last phrase indicates Anselm's appreciation of the cognitive irritation in having to respond that there is no further explanation. But the alternative would be to point to some prior difference between Satan and his angelic confrères, and that entails pushing the ultimate responsibility for Satan's fall back to God. Anselm cannot have that, so he must grant that there just is no more to be said.

This criticism involving the intelligibility of the libertarian choice finds expression in ancient Greek philosophy, a notable example, mentioned in Wikipedia, being the Stoic, Chryssipus. St. Augustine, the major philosophical influence on Anselm, brings it up against his Pelagian opponents.[1] Hume adds an interesting twist based on the ephemeral nature of actions. "[If libertarianism were true it would be the case that] Actions are by their very nature temporary and perishing... the person is not responsible for [an action]; and as it proceeded from nothing in him, that is durable or constant, and leaves nothing of that nature behind it, it is impossible he can, upon its account, become the object of punishment or vengeance."[2]

R. E. Hobart, in a 1934 paper in *Mind* provides the contemporary *locus classicus* for the question. (Peter van Inwagen refers to the problem as the "*Mind*" argument.)[3] Hobart argues that it is the person who is held responsible, but if the choice is not a manifestation of the person's character, then it seems we should not hold the person responsible for the choice. Indeterminists, says Hobart, ascribe free and undetermined actions to a "self." But when we, for example, reproach ourselves, "It is self we are reproaching, i.e., self that we are viewing as bad in that it produced bad actions. Except in so far as what-it-is produced these bad actions, there is no ground for reproaching it (calling it bad) and no meaning in doing so."[4]

In the last several decades many variants of the *Mind* argument have been presented and debated. In the next chapter, Chapter Eight, I attempt to tackle the luck problem head on by arguing that, given the Anselmian's concern for self-creation, Hume, Hobart, and their followers have gotten the relationship between choice and character backwards. Granted, having to allow that there is no explanation for S's ultimately choosing B over A—It was up to him and he just chose it!—is uncomfortable. But what we hold him responsible for is not the character that led to the choice, but the character *produced by* the choice. In the present chapter, in order to set the stage for this basic argument against the standard version of the luck problem, I discuss several preliminary issues.

[1] Rogers (2008): 36.

[2] David Hume, *A Treatise of Human Nature* Part III, Section II.

[3] Peter van Inwagen, *An Essay on Free Will* (Oxford: Clarendon Press, 1983): 126–50. van Inwagen (2000) allows—in a paper of the same name—that "Free Will Remains a Mystery."

[4] R. E. Hobart, "Free Will as Involving Determination and Inconceivable Without It." *Mind* 43 (1934): 1–27, at 4.

First it is useful to ask how the libertarian free choice is to be described. The free will literature is rife with conclusions drawn from intuitions, but that means that rhetoric—how the case is framed—may be unduly coloring the intuitions. Of course any case has to be framed *somehow*, and it would be silly to ask a philosopher not to express himself in the way most sympathetic to the point he is trying to make. I do not criticize participants in the free will debate for their rhetoric. I simply point it out as part of my effort to weaken the anti-libertarian intuition.

It seems to me that the fondness many libertarians have for describing libertarian choices in terms that involve assigning numerical probabilities has strengthened the intuitive case that there is something lucky or chancy in libertarian choice. Thus, I devote a section of the present chapter to the claim that, at least on the Anselmian account where freedom is grounded in aseity, exact numerical probabilities *cannot* ("cannot" even theoretically) be assigned. True, we can say that either possible option has a probability of more than 0 and less than 1, but that is the best we can do.

Finally I look at a recent attempt to go beyond the traditional way of casting the luck problem from Alfred Mele's *Free Will and Luck*. (This work will play a significant role in the present chapter in that it is perhaps the most carefully developed recent statement of the luck problem.) Mele holds that the most forceful and persuasive way to frame the problem is in terms of possible worlds; it is just a matter of luck that the non-determined agent occupies the actual world in which he chooses one way, rather than a possible world in which he chooses another. Discussing John and John2 who occupy different possible worlds, he writes, "If John's effort to resist temptation fails where John2's effort succeeds, and there is nothing about the agents' powers, capacities, states of mind, moral character, and the like that explains this difference in outcome, then the difference really is just a matter of luck."[5] Some critics of libertarianism allow that agent-causation, though it suffers from other problems, may escape the luck problem.[6] Mele, however, insists that, once we have seen that the criticism ought to be cast in terms of possible worlds, the luck problem remains even on the most robust agent-causation in which the agent is said to control his choice. This is because the problem is not that the actual choice in the actual world is somehow "chancy," but that the *cross-world differences* generate the luck. And

[5] Alfred R. Mele, *Free Will and Luck* (Oxford: Oxford University Press, 2006): 75–6 (Quoting "Ultimate Responsibility and Dumb Luck." *Social Philosophy and Policy* 16 (1999): 274–93, at 280).

[6] See, for example, Derk Pereboom, "Hard Incompatibilism and its Rivals." *Philosophical Studies* 144 (2009): 21–33, at 21–2. Pereboom's criticism of agent-causation—which he takes to be a sort of causation unique to human free will—is that it conflicts with "our best physical theories."

since libertarianism requires alternative possibilities, the (strong) libertarian must allow for the differences.[7]

In the present chapter I attempt to clarify just what the possible worlds locution *adds* to the traditional luck problem. I conclude that, on the representationalist account of what a "possible world" is, using the possible worlds locution does not really strengthen the luck problem. On the more colorful Lewisian account, the freedom, aseity, and self-creation that the Anselmian is concerned about become incoherent. The Anselmian, then, utterly rejects the Lewisian account and finds the representationalist account non-threatening. The upshot is that Mele's attempt does not succeed in *adding* something which renders the luck problem stronger than in its original "*Mind*" version. But, of course, the original version is powerful, as I grant in Chapter Eight.

Describing the *A Se* Choice

First, it is well to remember that Anselmian *a se* choice does not fit comfortably with the standard description of libertarian choice adopted by Mele and many others, including some who self-identify as libertarians. A common understanding of a libertarian choice holds that a choice is free if the agent can choose other than he actually chooses given the same past and laws of nature. (Or, expressed a little differently, there is a possible world with the same past and laws of nature in which the agent chooses other than he chooses in the actual world.) This is inadequate according to the Anselmian since it leaves it open for the agent's choice to fit the description, but be necessitated causally, or externally non-causally (ENC), or both, by factors ultimately outside of the agent.[8] For example, suppose God, consistent with the past and the laws of nature, can choose to make S choose A at t2 or make S choose B at t2. Then, simultaneously with S's choice at t2, God causes S to choose B. S's choice is from God, not from S, and so it is not an *a se* choice, even though he could have chosen otherwise consistent with the

[7] Mele (2006) suggests a "soft" libertarianism which he holds to be less susceptible to the luck problem. His thought is that there is some indeterminacy in the options that occur to the agent before the relevant decision is made. Anselmian aseity requires that the indeterminacy be at the point where the agent actually makes the choice, so a discussion of Mele's proposal lies outside the present work. Here the luck problem must be confronted head on. It cannot be addressed through tinkering with what a choice entails.

[8] Hugh McCann exploits this description to argue that, though God may cause all your choices, you may be free in a libertarian sense ("Divine Sovereignty and the Freedom of the Will." *Faith and Philosophy* 12 (1995): 582–9). Contemporary Molinists hold that there are true counterfactuals concerning libertarian freedom, which make it ENC necessary that an agent make a particular choice (see Chapter Four).

past and the laws of nature. Or suppose a mad scientist who happens to be a libertarian free agent sends signals from the future to make S choose B at t2, though the mad scientist could have decided to make S choose A at t2. S's choice for B is not free on the Anselmian account, even though he could have chosen otherwise consistent with the past and the laws of nature. Or suppose that the truth of Molinist counterfactuals is not something "in the past," and there is a true counterfactual: "S chooses B at t2 [in the situation which includes the past and the laws of nature]." Then, on the Anselmian account, S's choice is ENC necessitated, though the Molinist insists that it is some sort of libertarian choice and so could have been otherwise—in some Molinist sense—consistent with the past and the laws of nature.

Furthermore, that the agent could have done otherwise consistent with the past may not be a necessary condition of *a se* choice, depending on what one takes the past to contain. On the isotemporalist view of time, it is true always and everywhere that if S chooses B at t2, S chooses B at t2. So it is true before the choice that S chooses B at t2. And, if Anselm's God exists, it is true to say, before the choice, that He foreknows the choice. On the Anselmian account it is S who makes it true that S chooses B at t2, and so the necessity involved is a consequent necessity which does not conflict with or undermine aseity. But still, one might argue that it is inconsistent with the past that S choose A at t2, since it is true in the past that S chooses B at t2.

And how one understands the laws of nature may allow that a free choice might actually occur in our universe, yet not be consistent with the laws of nature. Suppose one takes the "oomph" view of causation and holds that the laws of nature describe the actions of substances exerting active power to produce effects. And suppose that when all these substances are operating normally, all choices are causally determined. But suppose that God "suspends" the laws of nature locally—He might prevent some substance from exercising its power—but leaves the laws operant *almost* always and everywhere in the universe. Then perhaps an agent might confront open options in a way that is not consistent with the laws of nature, in the sense that, had God not suspended the laws locally, the agent could not have confronted genuinely open options. In this scenario the Anselmian can say that the agent acted with libertarian freedom, even though such freedom is not consistent with the laws of nature. (This is not Anselm's view. I mention it only to show the inadequacy of some standard definitions of libertarianism as it applies to the Anselmian position.)

If, on the other hand, a "law of nature" describes the course of events that happens without exception, then God does not "suspend" laws of nature. Whatever "usual course of events" got locally suspended wasn't a "law," and, by hypothesis, any event that happens is consistent with the "laws." Or perhaps one could say that, if there is an omnipotent God, suspension of the laws is consistent with there being laws of nature, such that an agent who confronts open options when the laws are suspended nevertheless acts "consistently" with the laws of nature. That is, there are these laws, they just happen not to be operant at the moment.[9] My point here is just that describing a free choice as one which could be otherwise consistent with the past and the laws of nature does not capture the core of what the Anselmian is interested in. What counts is that the agent chooses *a se*, which entails that he is in the torn condition confronting open options at the time of choosing such that it is absolutely up to him which option he chooses.

How the action of making a choice is described may strengthen anti-libertarian sentiment. Epicurus famously associated an ability to do otherwise with atoms sometimes inexplicably swerving. This swerve, whether engaged in by ancient atoms or contemporary electrons, surely does not seem the right spot to locate the indeterminism which is required for freedom and responsibility. If my atoms made me do it, then I do not choose with aseity, unless some particular swerve can somehow be identified with the agent's consciously opting for this over that. At this moment in the development of particle physics and neurobiology that seems an unlikely identification. The philosopher defending libertarianism can rejoice that physics requires a rejection of universal determinism, but invoking the indeterminacy of sub-atomic particles to help describe a free choice is likely to exacerbate the luck problem.[10] (More on this later in the chapter.)

A millennium and a half ago, St. Augustine characterized his libertarian opponent Julian as holding that a choice "begins to be, but it does not have any source. Or, what is even crazier, it was not, it is, and nevertheless it never began."[11] Hobart offers a similar description of a libertarian choice, "[the choice] does not issue from any concrete continuing self; it is born at the moment, of nothing, hence it expresses no quality; it bursts into being from no source."[12] Or again, Hobart writes that indeterminism makes of the agent, "the helpless subject of an act of will that he suddenly finds discharging itself within him, though not emanating from what he is or prefers."[13] But this description of the choice

[9] Anselm and the medieval Christian tradition would not embrace the thought that God might suspend the laws of nature. They preferred to hold that God acts in concert with the laws of nature in performing miracles on the thesis that God made secondary causes because He really likes them and wouldn't act against them. But most of the medievals understood the "laws" as a consequence of the natures of substances, not as mere constant conjunctions.

[10] Helen Steward's associating agency with animals making trivial "choices" may fuel the luck argument in a similar way; *A Metaphysics for Freedom* (Oxford: Oxford University Press, 2012).

[11] *Opus imperfectum contra Julianum* 5.56 (my translation).

[12] Hobart (1934): 6–7.

[13] Hobart (1934): 22.

"bursting" into being from nowhere, or "discharging itself within him," is very distant from the Anselmian picture of the agent per-willing one of his two mutually exclusive desires to the point where it becomes his intention.

Mele also paints the libertarian choice in a way which bears little resemblance to Anselmian *a se* choice. In the Introduction to his book on the luck problem Mele discusses how a libertarian goddess, Diana, might construct libertarian free agents who are capable of making a "considered judgment that it would be best to A straightaway." In order to render her agents free she, "designs her agents in such a way that, even though they have just made such a judgment, and even though the judgment persists in the absence of biological damage, they may decide contrary to it."[14] This description does not fit Anselmian choice. It would be bizarre to make the considered judgment that it is best to A straightaway and yet find oneself deciding to the contrary. Except in cases of extreme akrasia, that isn't what we do, and certainly shouldn't be the sort of options we would want to have. If this is how the libertarian choice is described it is no wonder that it looks to be a matter of luck. In Anselm's torn condition (TC) the agent recognizes two mutually exclusive and morally significant options as desirable. Someone who has made the considered judgment that it is best to A (rather than to B) straightaway is not in the torn condition which is the requisite precursor to an *a se* choice. Of course the agent in TC might recognize option A as best in some sense, but he would also consider option B as best in some other way. Anselm's Satan, for example, knows quite well that he ought not to choose against what God has commanded, and so he knows that maintaining justice is "best" in one way, nonetheless he dearly *wants* to pursue the inappropriate good, and hence judges it to be "best" in another way. If he were not actively struggling to pursue incommensurate desires he could not make an *a se* choice.

Immediately after this description, Mele explains how one of Diana's agents comes to make a choice. On the hypothesis that the agent judges it best to A, "the probability of a decision to A is very high. Larger probabilities get a correspondingly larger segment of a tiny indeterministic neural roulette wheel in the agent's head than do smaller probabilities. A tiny neural ball bounces along the wheel; its landing in a particular segment is the agent's making the corresponding decision."[15] If making a choice is basically like the ball in a roulette wheel landing

[14] Mele (2006): 8.

[15] Mele (2006): 8. Daniel Dennett suggests that libertarian free choice is rather like having an internal "answer box" with a randomizer instructing subjects to answer questions in different ways; *Brainstorm* (Ann Arbor, MI: Bradford Books, 1978): 290. He also likens libertarian choice to flipping a coin (295). Derk Pereboom, in criticizing Kane's event-causal view, asks the reader to imagine that a "randomizing manipulator who spins a dial" provides the indeterminism in the process of choice;

then it does look to be a matter of luck. Roulette, after all, is a game of chance. But the Anselmian does not need to allow this, or any, description which portrays (or analogizes) a choice as chance, a toss of a coin, a throw of the dice, or any other event with which we ordinarily associate dumb luck. It is a rhetorical device in the service of anti-libertarianism. Whatever neural goings on provide the physical base or accompaniment for it, the *a se* choice for B is the agent consciously per-willing B. It is not at all like roulette.

In Mele's defense, he is often addressing his remarks to Robert Kane's event-causal libertarianism, and the picture he paints is not unlike the way Kane sometimes describes a choice. Kane, hoping to square his theory of free will with contemporary science, goes (in my view) a bit too far when he tries to link a conscious choice with the indeterminate behavior of sub-atomic particles in the brain through the "butterfly effect"; a tiny cause *here* (the electron(s) bouncing this way rather than that way) initiates a chain of events through a complex system of neural networks producing a large effect *there* (the conscious choice). Kane writes that a choice occurs "when either of the competing networks 'wins' (or reaches an activation threshold, which amounts to choice)"[16] That description—including the part about one network "winning"—really does make the process sound rather like Mele's neural roulette wheel. The Anselmian is sanguine that *a se* choice is consistent with the realities of physical nature. Perhaps, in the future, when neural science is more advanced, we will be able to isolate and observe the neural processes involved. The Anselmian predicts that they will not look much like the roulette wheel implanted by goddess Diana.

Kane's description of the process of choice seems especially susceptible to the luck criticism. Even his fellow libertarians of an agent-causal stripe bring it up against him. But Mele argues that the agent-causalists are not really in a better position. The bottom line, says Mele, is that even if we give the agent-causalist everything he wants in terms of agent control, the problem of luck remains. If there is a possible world with the same past and laws of nature in which the agent could have chosen other than he chose in our actual world, then it is just a matter of luck that the agent is in the actual world and not in that other possible world. I address this point later in the chapter, but there is another preliminary issue

Living Without Free Will (Cambridge: Cambridge University Press, 2001): 52. Pereboom grants that agent-causation, though it has other problems, does succeed in answering what he here calls the "Humean challenge."

[16] Robert Kane, "Libertarianism." In *Four Views on Free Will* edited by John Martin Fischer, Robert Kane, Derk Pereboom, and Manuel Vargas (Oxford: Blackwell, 2007): 5–43, at 28–9.

related to how we describe the libertarian free choice that needs to be considered, and that is the question of probabilities.

When I say "probabilities," I am speaking of what can be called "scientific" or "numerical" probabilities, not of our vague, everyday-language use of terms like "probable" and "probably." In connection with describing the act of choice, it seems to me that talking about probabilities has done a rhetorical disservice to libertarians. It is common to associate probabilities with inherently chancy sorts of events, like the bouncing of electrons, or with lucky happenings, like the landing of the ball on this or that number on the roulette wheel. In the contemporary literature, libertarians of every variety use probability language. Some just mention the probabilities, but others seem to take the assigning of numerical probabilities to be an important part of the project. Of course, at present, everyone recognizes that the thought that we could assign exact probabilities to the various options in a choice is just a fond hope. (Certainly we can say that the probability of the choice for some option is more than 0 and less than 1, but that is just another way of saying that it's possible that that option be chosen. That is not "assigning a probability" in some meaningful sense.) I will argue that on Anselmian libertarianism, and probably for any libertarianism which takes something like aseity to be a key factor, the issue is not just that assigning probabilities is presently scientifically impossible. Rather, assigning probabilities is impossible *simpliciter*. This is an intrinsically interesting point to make, and also it should encourage libertarians to consider abandoning the probability talk. Speaking as if probabilities can be assigned to choices links those choices rhetorically to all those chancy events from which libertarians should hope to distinguish free choice. In the battle for intuitions regarding free will, the probability talk supplies ammunition to the opponent, and so it is worthwhile to show why the assumption that probabilities can be assigned to libertarian free choices is misguided.

Aseity and Probabilities

Some philosophers—Robert Kane and Mark Balaguer are prime examples—suggest that, if libertarianism is true, numerical probabilities can, in principle, be assigned for a non-determined choice going one way or another.[17] Balaguer,

[17] Mark Balaguer, *Free Will as an Open Scientific Problem* (Cambridge, MA: MIT Press, 2010): 76–83; Robert Kane, *The Significance of Free Will* (Oxford: Oxford University Press, 1996): 177. Kane and Balaguer offer the most detailed proposals, but libertarians often mention probabilities in passing. Clarke (2003: 205) notes that while nondeterministic causation is usually understood as

for example, is not committed to libertarianism, but holds that it is for science to determine whether or not libertarianism is true. If libertarianism should turn out to be true, then probabilities can be assigned to the options in a choice. He writes that when an agent consciously feels equally attracted to two options and must choose between them, an objective probability of 0.5 should be assigned to each one of the two being chosen if the non-conscious factors influencing the agent towards each option are (roughly) evenly weighted. If the non-conscious factors are more heavily weighted towards one option, the probability of the agent choosing that option rises, and the probability of the agent choosing the other decreases. If the non-conscious factors are powerful enough to ensure that one option "has to" be chosen, then the probability of that option being chosen is 1, and the choice is determined.

Assigning numerical probabilities provides the discussion of libertarianism with a rhetorical veneer of scientific quantifiability which may enhance libertarianism's appeal in some circles, but I will argue that the options in an Anselmian free choice do not admit of probability assignments, assuming those assignments are taken to have predictive power. As we will see later, there may be understandings of what is meant by "probability" such that probabilities can be assigned. But these will be anomalous applications of the concept of probabilities, and quite useless if what we were hoping for was a way to assess the likelihood of one option being chosen over the other. For example, God may know the strengths of the motivating desires of an agent in the torn condition. And if "probability" is taken to mean "propensity," and "propensity" is understood, in this context, simply as the strength of the motivating desires, God may be able to assign probabilities. It does not follow that God—*per impossibile* setting aside His eternal knowledge of what the agent actually chooses—would know which outcome to bet on.

If we are using "probability" in its more common sense of relative frequency, then no one, not even God, can assign probabilities to *a se* choices.[18] There is a

probabilistic—allowing for fixed numerical probabilities, at least in principle—it need not be thought to be so.

[18] If might be argued that choices ultimately map onto subatomic indeterminacies in the brain, and such indeterminacies—on standard quantum mechanics—are subject to determinate probabilities, so it must be theoretically possible to assign probabilities to choices in a libertarian system. Derk Pereboom (2001: 81–6) uses this claim to question agent-causal theories that hold that agent-causation can be squared with natural laws. The problem, as Pereboom sees it, is that the causal efficacy contributed by the agent ought to affect the probabilities, but then it is difficult to see how this can be reconciled with the probabilistic behavior of the particles in the brain. My view is that, if analyzing choice as ultimately rooted in subatomic indeterminacy entails assigning determinate probabilities to the competing options in a choice, then this is yet another reason to be dubious about the project of associating free choice with subatomic indeterminacy.

qualifier to this claim, in that we can imagine two conceivable scenarios (Rewind and a relative from Chapter Six, note 42) which allow God to determine the relative frequency of one option over the other. But in these bizarre cases the assigning of probabilities will be "backward-looking" and—at least on the Anselmian view—devoid of predictive power. I believe that the argument I make here in the context of setting out Anselmian libertarianism probably applies to many other strongly libertarian theories. It lies outside the scope of the present work to defend this blanket claim, but I believe it is safe to say that probability talk adds nothing to libertarianism except for that (unearned because purely hypothetical) patina of scientific quantifiability.[19] But probability talk may undermine libertarianism rhetorically in that it strengthens the intuition that libertarian choice involves responsibility-denying luck. That being the case, libertarians should probably abandon it. More strongly, libertarians, like the Anselmian, who embrace the aseity requirement and the grounding principle entailed by it, simply cannot entertain the hope that we can assign probabilities to choices.

To see this, it is helpful to review the Anselmian schema for a libertarian free choice: At t1 agent S is in the torn condition, TC. He is debating between choosing A and choosing B, where he cannot do both, and doing one is morally preferable to doing the other. At t2 agent S chooses B. The choice for B over A is not causally or ENC necessitated by anything at all, and it is absolutely true to say, at t2 and thereafter, that S could have chosen A at t2.[20] In that it is up to S at t2 which option is chosen, the only possible grounding for the truth of propositions about a free choice, and the only possible source of knowledge about a free choice, is the actual agent's making the actual choice. This is the grounding principle as set out in Chapter Four.[21]

[19] In reviewing Balaguer's book, Gary Malinas writes, "In the absence of an account of dispositions, their strengths, and how to best represent and compare them, talk of brute probabilities of closing options and probabilistic causation are rhetorical devices that suggest a false precision in Balaguer's scientistic image of free will." *Analysis* 70 (2010): 793–5, at 793.

[20] In this context it is important to remember, as set out in Chapter Three, that the Anselmian choice is different from the libertarian choice as described by Robert Kane. Kane describes libertarianism as involving an indeterminate process which *precedes* the choice. He speaks of an effort of will being an, "*indeterminate* (event or process), thereby making the choice that terminates it *undetermined* [Kane's italics]" (Kane (1996: 128). The Anselmian schema locates the "moment" of indeterminism *at* the choice.

[21] Kane and Balaguer may complain that in the discussion which follows I operate at the macro level of desires, rather than the micro level of subatomic indeterminacies. I counter that if the probabilities in question are tied to the behavior of the particles, *rather than* to the conscious events within the agent, then they are not relevant to *a se* choice. An *a se* choice is a conscious event caused by a rational agent. If such an event can also be analyzed as subatomic, atomic, and neuronal activity, well and good. But if the probabilities are supposed to refer to the agent's future choice, then they must apply at least to the conscious event, whatever else they apply to.

Can prior probabilities be assigned to S's choice for A vs. his choice for B? At t2 S chooses B, so at t2 the "probability" of S choosing B is 1. Before t1 the conditions for S choosing are not in place, so let us focus on t1. The discussion can proceed more clearly if we imagine two libertarian philosophers, Black and Brown, who believe that probabilities can be assigned at t1, but who disagree about what the assignment should be. Black finds Balaguer's assessment intuitively plausible. If S is torn between A and B, and will choose in a non-determined way, then, if the conscious and non-conscious factors are equally weighted, the probability for each choice is 0.5. But, as Balaguer supposes, couldn't there be non-conscious factors present at t1 which "weight" the choice more towards B? That seems a plausible situation to Black, and, in that case it seems to Black that the probability that S will choose A is less than that S will choose B. Black imagines a day when neuroscience may be able to assign a probability of 0.3 (for example) to A, and 0.7 to B.

But Brown has a different intuition. "I grant," she says, "that there may be unconscious factors at work in S at t1. And perhaps, in some loose and metaphorical way they 'weight' or 'incline' S towards B. Indeed, S may consciously *feel* that one option 'pulls' upon him more strongly. In cases of trying to resist temptation, that might be the standard experience. Nevertheless, we have agreed that S's choice is not causally or ENC necessitated. Though I allow that in some sense S may be 'pulled', non-consciously or consciously, towards B, he really could, in the final analysis, choose A. What that means is that A is every bit as much a live option as B. But that is just another way of saying that the probabilities are 0.5 in either case. That, at any rate, is my intuition."

Of course we currently have no empirical way of addressing the issue between Black and Brown. The questions here are purely conceptual. First, what do we *mean* when we assign probabilities to options in a choice? And second, is there any sort of evidence that could possibly settle the question between Black and Brown? As the argument progresses we will move beyond the epistemic limitations of human investigators and ask: What sort of evidence could possibly settle the question for an omnipotent and omniscient God (bracketing divine foreknowledge and assuming that God knows only all there is to know about the past and the present, and that there is no sense at all in which the future is presently "fixed")?

There are roughly two ways of thinking about objective probabilities.[22] Sometimes probability talk aims to describe a causal propensity for an event, or kind of

[22] The probability talk in the free will literature does not seem to be about epistemic probabilities, and I do not know how questions of epistemic probability would function in the present context.

event, to happen. This seems to be the more usual sense in the libertarian literature, although it is not entirely clear. In other venues the more common understanding of probabilities refers to the relative frequency of a kind of repeatable event. There is no consensus among philosophers about which understanding of probability, if either, is to be preferred, and each confronts difficulties. For our purposes we can say enough to show that neither is likely to prove fruitful in analyzing libertarian freedom.

Libertarians sometimes speak of the motivational factors at work in a free choice as causing the choice "probabilistically" rather than necessarily. Robert Kane does so in order to counter the criticism that the libertarian choice is simply uncaused or inexplicable.[23] That suggests the "propensity" understanding of probability, in that the hope is to provide something more in the way of an explanation than just a brute conjunction between the agent's history and his choice. So let us look at that analysis first.[24] It will turn out that the thought of "probabilistic causation" is problematic for the libertarian.

As an example of probabilistic causation, at the macro level where it is easier to grasp, take the example of the coin toss, but assume that we do not have a "fair" coin. A tiny and unobtrusive shaving of metal has been added to the profile on the "head" side, making that side weightier. It seems very natural to say that, in such a case, the likelihood of the coin landing heads is greater than its landing tails. We might, for example, say that, because of the additional metal, on the next toss, there is a 0.7 chance of the coin landing heads, and a 0.3 chance of its landing tails. Although it *might* land tails, there is a *causal propensity* for it to land heads. If, sure enough, it lands heads, we may say that it was probabilistically caused to do so. (There is a caveat below.) A standard way of expressing probabilistic causation would be that "*c* probabilistically causes *e* if *c* and *e* are distinct events that occur and *c*'s occurrence raises the chance of *e*'s

Balaguer, I take it, believes that Future Science will be able to discover objective probabilities. So even though I couch the discussion in terms of what we can know or what our evidence might be or what probabilities can be "assigned," I take it that the sort of probabilities discussed in the literature are objective probabilities.

[23] Robert Kane (1996: 174–6). Here Kane is especially concerned to defend the view that the agent's reasons contribute to explaining his choice, even where the choice is not determined. My criticism here is aimed only at Kane's use of probability talk. His arguments concerning reasons as explanations are plausible.

[24] Timothy O'Connor uses the term "propensity" and holds that there is "a definite objective probability of [an option's] occurrence within the range (0,1), and this probability varies continuously as the agent is impacted by internal and external influences." "Degrees of Freedom." *Philosophical Explorations* 12 (2009): 119–25, at 120.

occurrence."[25] The probabilistic cause in the coin toss case is the adding of the metal shaving which, we suppose, raises the chances of the effect, the coin landing heads. If we were betting folk, heads would be the best bet.

How would the propensities understanding of probabilistic causation function in the context of a free choice and with relevance to the debate between Black and Brown? At t1 S desires both A and B, and at t2 S chooses B. Say that Black and Brown agree that there are non-conscious or conscious factors in some sense "favoring" B. We can think of them as elements of, or aspects of, the desire for B. Call them "weights" in B. Black assesses the situation this way: At t1 the desire for B, including the weights in B, raises the probability of S choosing B from 0.5—where it would be *sans* weights—to 0.7 (for example). When S chooses B at t2, Black explains that the desire for B, including the weights in B, probabilistically caused S to choose B over A. Black finds this a helpful way to describe what happened in that it allows that S's choice for B admits of a causal explanation, albeit a probabilistic one.

Brown is unconvinced. "As libertarians we agreed," she counters, "that S truly might have chosen A. And what would you say in that case? Your hypothesis, Black, was that the weights in B raise the probability of S choosing B to 0.7, and that is what the probabilistic causing entails. If S chooses A, as he surely might have done, then you must say that his choice for A was not probabilistically caused at all and that he chose *against* the option which he had a probabilistic cause to choose in favor of the option which he had no probabilistic cause to choose. That is incoherent!" Brown finds this an unacceptable conclusion. If invoking "probabilistic" causation was intended to shield the libertarian from the criticism that the choice is uncaused, it seems to do only half the job.[26] For any pair of options confronting an agent in TC, if the choice for one is probabilistically caused because there is a greater probability of it being chosen due to its "weights," then there will be a probabilistic cause for choosing one of the options and not the other. And in this case, half the job is worse than none. If there is, at

[25] Barry Loewer, "Freedom from Physics: Quantum Mechanics and Free Will." *Philosophical Topics* 24 (1996): 91–112, at 104. Loewer notes here that he is following Lewis' account, but adds that this is a very standard way of putting the principle. Richard Double, in criticizing Kane on the probabilistic causes of free choices, proposes a principle, "(PRE): Citing a person's deliberative process P rationally explains a choice C only if the probability of C given P is greater than the probability of not C given P." *The Non-reality of Free Will* (Oxford: Oxford University Press, 1991): 203–4. As Kane notes (1996: 176) this principle can be undermined by counterexamples where some probabilistic cause P *raises* the probability of some effect C, but does not render the effect C more likely than not. But PRE seems a non-standard way of stating probabilistic causation, so Kane's dismissal of PRE does not succeed in addressing the problem with probabilities.

[26] Richard Double makes a somewhat similar point in "Libertarianism and Rationality." *The Southern Journal of Philosophy* 26 (1988): 431–9.

t1, a cause—albeit probabilistic—for S's choosing B, but no cause for S's choosing A, then A does not seem to be an equally viable option to B and so, at t1, it does not seem that S faces robust alternatives.

With our coin which was weighted for heads, should it happen to come up tails, we would probably assume that some causal factor had been at work, counteracting the added shaving, to produce the less likely result. That is because we do not suppose that a coin toss is *really* not determined by preceding factors. We assume that the coin toss system is a macro system which is too complex to allow us to make accurate predictions, but not so complex that micro indeterminacy could translate into it being *genuinely* not determined which side of the coin lands up. God, without knowing the future, but by knowing everything about the coin toss system prior to the landing of the coin, can know which way the coin will come up.[27] Still, from our perspective, it appears undetermined, probability talk is appropriate, and referring to the addition of the shaving as "probabilisticially causing" the coin to come up heads is the correct locution. But the libertarian thesis is that S's choice is truly not determined. That entails that S can opt for A as well as for B, and that seems problematic on Black's analysis of probabilistic causation.

Brown, while she may reject Black's understanding of probabilities, might try to reconcile probabilistic causation with her intuition that at t1 the probabilities of S choosing B and of S choosing A are each 0.5. Recall the statement of the principle above, "*c* probabilistically causes *e* if *c* and *e* are distinct events that occur and *c*'s occurrence raises the chance of *e*'s occurrence." Brown may hold that S could not make the choice for B over A at t2 if S were not in TC between A and B at t1. If—Brown may argue—S did not desire B at t1 then the probability of S choosing B over A at t2 would be 0. S's desiring B at t1 raises the probability of S choosing B at t2, and so S's desiring B at t1 is a probabilistic cause of S's choosing B at t2. But, continues Brown, S could not *choose* B *over A* if S did not desire A at t1. That means that S's desiring A at t1 raises the probabilities of S choosing B over A at t2 above what they would be—that is 0—if S did not desire A. And so, since the desires for A and for B both raise the probability of S choosing B over A above 0, which is where it would be in the absence of either, and since they are equally necessary for S's choice, *and* since our description of what a choice is entails robust alternatives, the desires for A and for B should each be said to produce a probability of 0.5 of S's choosing A and a probability of 0.5 of S's choosing B.

[27] I take it that this means that, in the coin-toss case, objective—in the eyes of God—probability, and epistemic—quoad nos—probability come apart. God knows which way the coin will land, since in fact it is causally necessitated, so the objective prior probability of heads is either 1 or 0.

But Brown has gotten herself into a muddle. Given the proposed description of a free choice, she is correct to claim that S's desiring A and S's desiring B are both required in order for S to choose B over A, but her conclusion about probabilistic causation is surely odd. She seems to be saying that S's desire for B probabilistically caused his choosing B over A, *and* that S's desire for A probabilistically caused his choosing B over A. That can't be right. For one thing, the example stipulates that A and B are mutually exclusive options. Pursuing the desire for A to the point of choosing A rules out the possibility of choosing B, and pursuing the desire for B to the point of choosing B rules out the possibility of choosing A. It seems bizarre to claim that the desires for both equally probabilistically *cause* the choice for one over the other. (The very phrase "equally probabilistically cause," in this context, seems contrary to the thought that probabilistic causation refers to what raises the probabilities of an effect.) Moreover, had S chosen A over B, Brown would have to say that the same probabilistic causes produced *that* choice. But then the appeal to probabilistic causes has no explanatory value in regard to the choice actually made.

Note that the problem here is not that Black and Brown simply lack the evidence for determining which of them is right about how to assign probabilities to S's choice. They have not succeeded in making sense of assigning probabilities, where the probability is supposed to reflect some causal tendency or propensity. Moreover, there is absolutely no way for human science to resolve the difference between them regarding their intuitions on assigning probabilities. Suppose (as is probably impossible) science could one day monitor the various conscious and non-conscious elements at work in S's desires as he is in TC and assign some sort of "weights" to them. The question between Brown and Black is whether those weights budge the probability of S choosing B off of 0.5. Merely recording and assigning weights to the desires does not settle that question, unless we take the assignment of probabilities simply to be the assigning of the weights. If assigning probabilities is supposed to let us judge, at t1, what S is more likely to choose at t2, simply observing S's condition at t1 does not provide either Black or Brown with evidence to support their side of the debate.

The underlying issue here can be seen more clearly if we bypass the human observer, who is subject to ineradicable limitations, and say that it is God, in His omniscience, Who is observing and "weighting" the strengths of S's desires in TC. The hypothesis is that there are weights in B which, in some sense, favor B. Maybe S *feels* pulled towards B. But it is also the case that S is genuinely in TC. God may be able to assign probabilities in the sense that He may be able to note the strengths of the desires in S in TC at t1. But, unlike with Robert Kane's theory, it is not these preceding desires (or "efforts" in Kane's system) that produce the choice for this over that. The choice is up to the agent. Knowing all there is to know about the preceding desires does not provide information which allows one—even God—to predict how the agent will choose, or even to assess what the agent is *more likely* to choose. Since the case of S choosing between A and B is absolutely unique, and the preceding desires do not cause the choice, the only information concerning what S *will* choose comes from what S *does* choose. Just as even God cannot know "before" S chooses what S will choose, even God cannot assess the *likelihood* of S choosing one way or the other.

The only evidence to which human science has access regarding propensities must come from observing repeated instances of some cause producing some effect. In this case science would have to observe repeated instances of S being in TC at t1, then choosing A or B at t2. This means that in trying to decide whether or not the "weights" in question affect probabilities, we must appeal to the more common understanding of probability talk, relative frequency of a repeatable event. We might suppose that the relative frequency is a function of inherent propensities, but, absent repeated trials, these propensities are invisible to us. So let us turn to the relative frequency understanding of probabilities and ask whether the options in a free choice are amenable to probability assignments on this approach.

Back to our weighted coin. Say we toss it 1,000 times. It comes up heads 711 times and tails 289 times. On the relative frequency understanding of probability, what we mean when we say that on the next toss, number 1001, the coin has a probability of (roughly) 0.7 of coming up heads, is just that, in repeated trials, the coin has come up heads (roughly) 70 percent of the tosses. Analysis of the indeterminate behavior of sub-atomic particles lends itself to this sort of probability talk, since one electron is much like another, and the sorts of situations in which electrons find themselves are reproducible. But the problem with attempting to associate this sort of probability with the non-determined alternatives in libertarian free choice is obvious. We are talking about individual human agents in individual situations. (Indeed, the issue of singular events is one motivation for the "propensities" understanding of probabilities. On the relative frequency view, if no one ever tossed our weighted coin, it would not have a probability of landing either way! Conversely, as evidenced by our having to move to the relative frequency approach, it is difficult to talk about a propensity without invoking relative frequency, and some—though not the present author—suspect that propensity talk is really frequency talk in disguise.) At least prima facie, for any libertarian free agent S, S's choice seems to be about the *least* repeatable sort of event possible.[28] S is a unique individual, unlike any other. And S's being in TC between A and B at t1 is a unique situation. True, S may be torn between alternatives very like A and B more than once in his life, but each time will be a different situation. For example, the second instance will involve a history of S's having made a past choice for A (or something A-like) or B (or something B-like), which makes the second instance very different from the first.

Is there any conceivable scenario in which probability, understood as relative frequency, can be assigned to the options in a future choice? Yes, actually, there are two (mentioned as "Rewind" and "Copier" in Chapter Six), but they will have to include an omnipotent, omniscient God. Peter van Inwagen suggests a case where an agent makes a libertarian free choice, and then God rewinds the universe to a point before the choice, and the agent makes the choice again. God does this 1,000 times. In van Inwagen's example, the agent chooses each way roughly half the time, and so the next iteration of the choice can be assigned a probability of 0.5.[29] (Perhaps the fact that van Inwagen chooses 0.5 as the probability in his example indicates an intuitive sympathy for Brown's side of the Black/Brown debate.) But note that God *actually* has to rewind the universe 1,000 times. It will not do to ask how the agent *would* choose if God *were to* rewind the universe 1,000 times. There is no answer to that question. The grounding principle entails that the truth of propositions about a free choice, and the source of knowledge about a free choice, must come from the agent actually making the choice. If God does not actually rewind the universe a sufficient number of times, then there is no probability, where probability is understood as relative frequency. A similar scenario involves multiple universes. What if God, at t1, "copies" the universe so that there are 1,000 iterations of the universe at t1, including S in TC. (For the sake of argument bracket the problem of the identity of the multiple S's.) God can then watch to see how many of the 1,000 S's choose A and how many choose B at t2, and then He can rewind one of the universes to t1 and assign a probability to S's future choice in that universe.

[28] Peter van Inwagen mentions the issue of describing singular events as "random" or "chance." *An Essay on Free Will* (Oxford: Clarendon Press, 1983): 128–9. Ephraim Suhir, writing a text book on *Applied Probability for Engineers and Scientists* (New York: McGraw-Hill, 1997) explains that the probabilistic approach is applied to situations involving "products manufactured in large quantities... [and]... experiments, which are repeated many times in identical conditions... [such methods]... cannot be applied in situations where the conditions of an experiment or a trial are not reproducible or when the events are very rare." (xx–xxi). Derk Pereboom (2001: 83–5), writes that some agent-causalists propose that "the frequencies of agent-caused free choices dovetail with determinate physical probabilities," but I simply do not know what "frequencies" can mean here.

[29] Peter van Inwagen (2000): 13–16. van Inwagen offers this example as support for the criticism that libertarian free choice happens "by chance."

But again, there can be a justified probability assignment only if God actually does produce the 1,000 copies of the universe at t1. There is no answer to the question of how many times S *would* choose A if God *were* to copy the universe 1,000 times.

On the relative frequency understanding of probability, unless God actually does rewind or copy the universe, the options in future choices cannot be given probability assignments. This is not an epistemic problem. Without the rewinding or copying there *are no objective probabilities* to be assigned. For those who do not believe in God it is even harder to imagine scenarios in which libertarian free choices could admit of justified probability assignments. (Could the universe get rewound or copied 1,000 times by natural causes?) For those who believe in God there is no reason to think He engages in such shenanigans and every reason to believe He does not. (As noted in Chapter Six, a paramount reason is that rewinding or copying would involve systematic deception.)

Perhaps most importantly, on Rewind and Copier, although God can now assign a probability to S's "next" choice between A and B, all that probability refers to is the relative frequencies in the 1,000 instances. With a non-determined, *a se* choice it is not clear how these relative frequencies are to be translated into any helpful *predictive* power—if that is what we are after. As the Anselmian has described it, an *a se* choice is not brought about by preceding causes. It is absolutely up to the agent. So even if we note causal propensities and even if we record how things go with the 1,000 choices, it is not clear that we (or God) can translate that into useful information about how the next instance of the choice is likely to go.

Suppose that in Rewind or Copier S chooses B 711 times. Black may say that this means that the *next* time S is in TC at t1 there is a 0.7 chance of S choosing B at t2, and so we should bet on S choosing B. But Brown may still assert that—whatever the statistics in the 1,000 instances—in that A and B are both viable options, there is no more likelihood of one outcome than the other. She may insist that, were God to reproduce Rewind or Copier and engage in *another* set of 1,000 instances of t1, the relative frequency of A and B could be radically different from the original set of 1,000. We may use the term "likelihood," but if it just *means* the relative frequency, then we are not helped in our attempt to predict. Human choices are nothing like coin tosses and bouncing particles, and it is best to jettison talk of assigning scientific or numerical probabilities to options in future choices.

It does not follow that we must do violence to our intuitions or to common ways of speaking about choices. We may recognize conscious or non-conscious "weights" at work when people make choices. Say that S in TC is very tempted towards some morally wrong A and really has to exert himself to fight against the temptation in order to choose the better B. Based on our knowing about the temptation we might well predict that he will choose A and then be surprised when he chooses B. Had we been quite clear that he was in TC we might be less surprised, but it may be impossible for third parties to tell when someone is in TC... even the agent himself may not clearly grasp what his current situation is relative to an impending choice. And jettisoning the probability talk does not entail that we should not praise S for the very fact that he chose B when he was so strongly drawn towards A. Business as usual is in no way enhanced by assigning probabilities. Some libertarians might wish to retain talk of numerical probabilities in the belief that it lends the discussion a scientific tone, but, as Aristotle said, we must not ask for more exactness than our topic allows. We smile today at Bentham's attempt to quantify pleasure and pain and produce a moral calculus. We recognize his failed effort as symptomatic of that naïve, nineteenth-century faith in the power of Science to address and solve all problems. The study of brains at work is likely to produce all sorts of exciting and useful knowledge related to human choosing. But robustly libertarian free choices, if such there be, are unlikely to ever become the subjects of justified probability assignments. Probability talk on the part of libertarians plays into the critic's claim that libertarian choice involves responsibility-denying luck, and so it is better abandoned.

Mele's Possible Worlds Argument

Having addressed the question of rhetoric and intuitions vis-à-vis the luck problem, it is time to turn to a recent version of the problem.[30] As I noted in the introduction, the luck criticism of libertarianism has been around since St. Augustine's day in one form or another, and I will address the more "classical" statement of the problem as it appears in Hume and in the discussion in *Mind* from the mid twentieth century in Chapter Eight. Here I want to look at the "possible worlds" twist on this old criticism which has been proposed by Alfred Mele. My question is: Does the possible worlds locution actually recast the problem significantly or contribute some new element to it such that the defender of libertarianism now confronts *two* luck problems, the original and the possible worlds version? I will conclude that Mele's possible worlds approach does not ultimately add anything to the older way of putting the criticism. This conclusion leaves the original luck problem still standing strong, but at least the Anselmian

[30] Thanks to Michael Rea for helpful comments on this last section of Chapter 7.

has only the one, original problem to address. I tackle the classical version of the luck problem in Chapter Eight.

Some contemporary participants in the free will debate have argued that, as far as the luck problem is concerned, agent-causalists are in a better position than event-causalists. Mele disagrees. He holds that the possible worlds statement of the problem is as effective against agent-causalists as against event-causalists. One striking aspect of Mele's version of the luck problem is that he takes the novel approach of granting agent-causalists everything they have proposed as requisite for a robustly free choice. On a well-developed agent-causal view the problem, according to Mele, is not that there is something lucky or chancy about the actual choice as it is made in our actual world. If S makes a libertarian choice for B in the actual world, Mele allows that—given certain agent-causal criteria—S's choice for B is not a matter of luck. This means that Mele need not admit defeat when confronted with the plausible claim, which libertarians make in their defense, that when some agent deliberately does something it cannot be considered merely a matter of luck.[31] True, says Mele. S's actual choice is not a matter of luck. But the problem, in Mele's view, is that this does not answer the difficulty of "cross-world luck."[32] On the libertarian account, if S chooses B at t2 in our actual world, there is a possible world with the same past and laws of nature in which S chooses A at t2. Discussing O'Connor's agent-causal view Mele writes:

> Even if the fact that Tim exercised direct control in choosing to continue working is incompatible with its being just a matter of luck that he chose to continue working, this does not show that a relevant cross-world difference between his exercising direct control "in [this] particular way"... and his exercising it in choosing to do something else is not just a matter of luck.[33]

And against Clarke:

> As far as I can see, even if Ann's exercise of direct active control in *W* [our actual world] in tossing the coin is a *full-blown*—that is, undiminished and unreduced...—exercise of such control and the same is true of her exercise of direct active control in *Wn* [another possible world with the same past and laws of nature as *W*] in deciding not to toss it, the difference between *W* and *Wn* in exercises of direct active control at noon is just a matter of luck.[34]

[31] Meghan Griffith, "Why Agent-Causal Actions Are not Lucky." *American Philosophical Quarterly* 47 (2010): 43–56. In a recent article ("Moral Responsibility and the Continuation Problem." *Philosophical Studies* 162 (2013): 237–55), Mele changes the labeling. Rather than refer to a "luck" problem, which led many critics to ponder the meaning and nature of "luck," the problem is that libertarianism insists on possible worlds allowing for different "continuations" depending on the agent's actual choice versus non-actual possible choices. I do not see any substantive difference in Mele's point. The article is aimed at event-causationists in any case.

[32] Mele (2006): 49–80. [33] Mele (2006): 55. [34] Mele (2006): 65.

What does Mele mean by "luck"?

Well, if the question why an agent exercised his agent-causal power at *t* in deciding to *A* rather than exercising it at *t* in any of the alternative ways he does in other possible worlds with the same past and laws of nature is, in principle, unanswerable . . . and his exercising it at *t* in so deciding has an effect on how his life goes, I count that as luck for the agent.[35]

In a grumpy mood the libertarian may respond that Mele is just begging the question. If Mele's understanding of responsibility-undermining "luck" just is, or is an entailment of, the libertarian's requirement that the agent be able to do otherwise, then, one might argue, it is hard to see what Mele has added to the discussion. This might seem especially the case in that, as noted above, Mele allows the agent-causalist's claim to his various control-enhancing, agent-causal properties. Mele, however, takes his focus on cross-world differences to be a new element in the discussion and one that ought to strengthen the intuition that libertarian accounts involve the sort of luck that undermines responsibility.

I take it that there has been something of an intuitive stalemate in the literature regarding whether or not the luck problem succeeds in undermining libertarianism. Mele's focus on cross-world differences constitutes a new effort to nudge the debate forward by enhancing the anti-libertarian intuition. But just what is it about the possible worlds version of the problem that supports Mele's intuition concerning libertarian freedom? What does the possible worlds talk *add* to the earlier expressions of the problem, such as those found in Hume or in the *Mind* debate of several decades ago? To my knowledge, although most parties to the free will debate seem comfortable talking about possible worlds, no one has tried to set out just what this talk means in the context. I allow that appeal to *some* possible worlds version of the problem (perhaps on some non-standard understanding of possible worlds?) might possibly budge intuitions in the direction of rejecting libertarianism. However, I will try to show at least this; *Mele's* way of putting the problem cannot add anything to the original luck problem vis-à-vis Anselmian libertarianism. If the Anselmian response to the original problem in the next chapter is successful, then the luck problem does not pose an insurmountable obstacle to accepting Anselm's theory.

I argued in Chapters Three and Six that the ability to choose otherwise is central to Anselmian libertarianism. So if, as in our standard example, in our actual world S chooses B at t2, the Anselmian must say it was possible for S to choose A at t2. What he means is that, at t1 S was in the torn condition debating between choosing A and choosing B and that his choice for B was *a se*, which

[35] Mele (2006): 70.

entails that it was not causally or ENC necessitated. "It was possible for S to choose A at t2" is a statement about what is the case in the actual world.[36] So far, for someone who was not moved by the classical statement of the luck problem, there does not seem to be anything in the admission of this possibility to *enhance* the intuition that libertarianism undermines responsibility.

Suppose we recast "It was possible for S to choose A at t2" as "There is a possible world in which S chooses A at t2." Call our actual world W. In W S chooses B at t2. And let's talk about only one other possible world, W*, in which S chooses A at t2. To my knowledge, there are two main ways to understand possible worlds talk. Let us look first at actualist representationalism (AR). AR holds that "a possible world is an actual maximally consistent representation of how the universe could possibly have been, and the actual world is the representation of how the universe actually is."[37] The "actual world" is a possible world, so it is a representation of what actually exists. The "universe" is what exists. So existence in the actual world is reducible to existence *simpliciter*.

If this is Mele's understanding of what possible worlds are, it is difficult to see where the possible worlds talk *enhances* the anti-libertarian intuition. "There is a possible world (W*) in which S chooses A at t2" says that, in W there is a consistent representation (W*) which includes all the propositions (if our representation is propositional) about the past and the laws of nature and also "S chooses A at t2." But W* is only a representation existing within W. The most interesting cross-world difference between W and W* is that W represents *reality* and W* does not. Is it just luck that W is reality and W* is not? Well, it was S's free choice which made it the case that W represents the existent universe, while W* is the mere representation within it. Mele has allowed that the agent may have robust control over the actual choice, and so the agent's choice in the actual world is not lucky. If the actual choice is not lucky, the actual choice is the only choice that exists, and it is the choice which brings about W's representing the existing universe and W*'s not, it is hard to see where luck comes into play. "Ah, but in W we can consistently represent W* in which S chooses A," Mele might say. Well, that certainly is what libertarians hold, but I would find it deeply puzzling if Mele believed that this contributed something *new* to the undermining of libertarianism. True, that we can consistently represent W* entails that there is nothing about S's past to explain S's ultimate preference for B over A. But the claim that, if the choice is not mere luck there *must* be something in S's past to

[36] A similar point was made about counterfactual statements in Chapter Four.

[37] I am following Takashi Yagisawa (2009) in the *Stanford Encyclopedia of Philosophy* under "Possible Objects."

explain the preference, seems to be the same old argument from Hobart and Hume and Augustine to which I will turn in Chapter Eight.

Before turning to the other standard way of understanding the possible worlds locution we should look at a hypothesis which is not exactly a possible worlds view, but which is similar and is sometimes suggested by Mele's way of expressing the issue; the hypothesis of a "branching" universe. Mele writes that, on libertarianism, W and W* "do not diverge before the agent decides." The term "diverge" could suggest something like "to move or extend in different directions from a common point." This, and the phrasing that there is "*the* agent," only one agent, could suggest the thought that we live in a branching universe. At each moment where more than one possibility can be actualized, whatever possibilities can be actualized are actualized.[38] Whether or not Mele is allowing the possibility of a branching universe, some philosophers do, so it is worthwhile to ask how Anselmian libertarianism fares in the face of a luck criticism rooted in the thought of a branching universe.

On the branching hypothesis a choice might look like this: If, at t1, it is really possible that S choose A or B, then at t2 the universe branches and S chooses A *and* B. From the original stem at t1, W and W* both emerge, and which is actual is indexed to the perceiver in W or W*. And I suppose we should say that on the branching universe view S bifurcates into S' and S* each of whom go their merry way in W or W* as the case may be. We need the prime with S' to show that this S is not somehow more continuous with the "original" S at t1 than is S*. Mele very often speaks as if it is the identical agent in W and in W*. For example, he writes, "To be sure, something about Joe may explain why it is *possible* for him to decide to *A* in the actual world and decide not to *A* in another world with the same laws and past."[39] "Joe" in the sentence seems to refer to one individual person. Can there be two "instances" of the same individual person in different (really existent) branches? One individual person engaged in different, mutually exclusive actions? In a branching universe it would certainly get confusing if S *simpliciter* chose A and S *simpliciter* chose B (leaving aside all the quantum

[38] Some interpret quantum mechanics as entailing a "many-worlds" theory (like my "branching" worlds theory, but located within the discipline of physics). David Hodgson argues that the many-worlds theory is absurd. For one thing, it renders the assigning of probabilities impossible; "Quantum Physics, Consciousness, and Free Will." In *The Oxford Handbook of Free Will*, First Edition, edited by Robert Kane (Oxford: Oxford University Press, 2002): 85–110. He writes that on the many-worlds theory, "the probabilities of [quantum mechanics] cannot be expressed in the statistics of actual occurrences and nonoccurences *because there are never any nonoccurrences* [Hodgson's italics]. This makes nonsense of [quantum mechanics]" (98). I will argue that it also makes nonsense of libertarian free choice.

[39] Mele (2006): 8.

indeterminacies and the plethora of branches *they* would produce). Could S *simpliciter* be the sum of all the Ss on all the different branches? Happily, the Anselmian has ample reason to reject this branching hypothesis and so can leave it to others to try to make sense of it.

Prima facie all this branching seems to make for an unparsimonious multiverse of the sort that one ought to endorse only if it were absolutely necessary. But set that aside. The branching thesis conflicts with Anselmian libertarianism, and probably with other libertarian views as well, so the Anselmian has every reason to reject it. The metaphysical hypothesis is that all possibilities must be actualized. If S is in the torn condition at t1 then, assuming he goes on to make a choice, he must choose (or must bifurcate and choose) both A and B, bringing W and W* into being. That is, given the overall constitution of the branching universe, it is necessary that he choose A and it is necessary that he choose B. Our definition of *a se* choice included the thought that if there was an ENC necessity—some (non-causal) truth about reality, existing independently of the agent, but necessitating the agent's choice, then the choice is not *a se*. (See the discussion of Molinism in Chapter Four and Rewind in Chapter Six.) It is not a matter of luck that S' is in W and S* is in W*. It is a matter of necessity entailed by the branching universe. But then S's choice is not free. S cannot possibly fail to choose both A and B. The theist Anselmian can add that God, with His interest in our constructing our characters, would not set things up so that an agent in the torn condition must choose both the morally good and the morally wicked option. The branching hypothesis entails a morally senseless branching universe, and anyone endorsing libertarianism as the best theory for grounding moral responsibility will have nothing to do with it. If Mele's intuition concerning the luck involved in cross-world differences presupposes the branching view, then it is not a threat to Anselmian libertarians.

The other main possible worlds theory is possibilist realism (PR), associated especially with David Lewis. "When possibilist realists assert, 'Non-actual possible objects exist,' their word 'exist' has the same linguistic meaning as when actualists assert, 'Actual objects exist'... All realms of existence are metaphysically on a par with one another."[40] All possible worlds exist and "actual" is indexical to the world which includes the perceiver who is employing the term. PR is the sort of view that would be uncharitable to attribute to a fellow philosopher unless he had explicitly embraced it. Mele does not offer an explication of how he understands the possible worlds locution, but his phrasing does seem to suggest a sort of ontological parallelism between W and W*. How does

[40] Yagisawa (2009).

Mele's luck problem play out on the PR supposition that W and W* are equally existent and which world is "actual" is indexical to which world you are in?

Though Mele does not do so, it is usual, on this analysis, to speak of counterparts. (I suppose in the branching hypothesis one might label S* a counterpart of S' and vice versa.) In W there is S, and in W* there is counterpart-S (S*). W and W* "diverge," but not in the branching sense that they come from the same "stem" world. They are two possible worlds, causally unconnected, in which the past of W and the counterpart-past of W* (past*) and the laws of nature of W and the counterpart-laws of nature of W* (laws of nature*), are the "same" (that is, they are as similar as possible without W and W* being the same possible world) up through t1 and t1*. They "diverge" at t2 and t2* in the sense that S chooses B at t2 in W and S* chooses A at t2* in W*.

Again, set aside the question of parsimony. Given the way Mele phrases his luck problem he seems to hold that it is a matter of luck for S in W that he is not in W*. But on PR—where the multiple, actually existent worlds are causally separate—that is a metaphysical impossibility. The hypothesis is that there is no interaction between the discreet possible worlds. S can't possibly get to W* no matter what he does. Would the luck problem be better put if the claim is that S is lucky that W is the actual world relative to S and S* is lucky that W* is the actual world relative to S*? But, on the assumption that S and S* are libertarian free agents, it is S's actual choice that brings it about that W is his actuality and S*'s actual choice that brings it about that W* is his actuality. And Mele has granted to the agent-causalists that the element of luck that he is concerned with does *not* attach to the actual agents making the actual choices, but to the "cross-world differences." But in that, on PR, by Mele's admission, the luck does not attach to S's agent-caused choice, how is it a matter of luck *for S* that he is in W, since he himself brought it about by his *non-lucky* choosing. And the same with S* in W*.

But there is a further twist. The way Mele consistently sets up the luck problem, S's choosing B in W *entails* that there is W* in which S* chooses A. On the AR analysis of possible worlds this isn't terminally bizarre, since all that is being said is that in W we can consistently represent S's choosing other than he actually chooses. It isn't bizarre, but it doesn't help Mele. It is simply a way of stating the original version of the luck problem, and so it shouldn't add to or strengthen the anti-libertarian intuition beyond what the original luck problem could elicit. But on PR, Mele would seem to be saying that, on libertarianism, it is *necessary* that both W and W* exist, such that, if S chooses B in W, then necessarily, S* chooses A in W*—and vice versa! If B is the unhappy choice—and in spite of the fact that S had agent-causal control over his choice, so the choice itself was not a matter of luck—S might bemoan his bad luck in being in

W and wish he had been S* in W*. But is this coherent? Given that—as Mele expresses it—the possibility that S chooses B *must* be actualized in some (on PR actually existent) possible world, it is *necessary* that someone be S in W, and he is that someone. And so the same problem arises as with the branching universe. If S's choice is ENC necessitated because PR is true and all possibilities *must be* actualized at their own worlds, then it is not an *a se* choice, and we live in a morally senseless multiverse.

A part of my methodology has been to note that intuitions concerning free will may differ depending on one's background worldview, and so I have tried to point out where the Anselmian's background theism may affect her intuitions. Perhaps difficulty harmonizing classical theism with the PR understanding of the possible worlds locution colors my claim that Mele's new twist on the luck problem should not strengthen anyone's intuitions against libertarianism. It is often said that God, as a necessary being, must exist in all possible worlds. On the AR understanding of the possible worlds locution, that is not a problem. Any consistent representation of a world must include the existence of God. But on PR, the classical God's necessary existence cannot mean that each possible world is the product of its own divinity, since in that case each "god" would not be God. There are many avenues to this conclusion. The simplest is to note that the God of classical theism is unique. If there are more than one of X—even if the multiple Xs are not causally related—no X can be God. Another approach to the same conclusion is that God is omnipotent. If there is another divinity—in any world at all—over which He has no control, He is not omnipotent. The theist might hold that, if the existence of God conflicts with PR, then PR is not the case. That strikes me as the most reasonable conclusion. But for the sake of argument, if one felt compelled to retain something like PR, then one might adjust Lewis' description a bit and allow that the one God might produce the many worlds, but hold that they are causally unrelated to one another in terms of *secondary* causation. In that they are all produced by God they cannot be unrelated in terms of primary causation. They must overlap on God.

The Anselmian, of course, must reject this picture since it is morally senseless in much the same way that the branching universe is. The whole point in positing libertarian freedom, for the Anselmian, is to allow the agent to make *a se* choices and contribute to his own creation. Mele's way of describing the free choice vis-à-vis possible worlds, if he is assuming PR, certainly makes it sound as if some ENC necessity accrues to each "choice" in each possible world. So the Anselmian must suppose that this is not the way God sets things up.

If Mele's intuition concerning the luck involved in cross-world differences presupposes PR, then it is not a threat to Anselmian libertarians, who will just insist that reality is not as PR says it is. Perhaps Mele has a different theory of possible worlds in mind such that an understanding of that theory would bolster the luck problem. Or perhaps my reading of the AR and PR analyses of possible worlds misses the mark, and Mele can appeal to one of these theories to make his case. But there is still a difficulty. The question of how to analyze possible worlds is a metaphysical question. It is problematic to ask for intuitions on metaphysical questions. Mele introduces his possible worlds version of the luck problem to enhance its intuitive appeal. But presumably he is presupposing one or another theory regarding possible worlds. (If not, then what do his claims about possible worlds mean?) But it does not seem likely that some shared or helpful intuition will be forthcoming out of such a metaphysically laden (if not perfectly clear) formulation of the luck problem. And if casting the problem in terms of possible worlds does not add anything substantive to the criticism, then the luck problem seems to resolve into its original form; in the absence of anything in the agent's past to explain his choosing one way over another it seems that that agent per se should not be held responsible. So let us return to the original problem as set out by Hume and Hobart, and recognized by Anselm himself. That will be the topic of Chapter Eight.

8

The Luck Problem

Part II. The Locus of Responsibility

> So just as [after Satan's fall] the bad angel is to be blamed because he is not able to return to justice, so [the good angel] is to be praised because he cannot abandon it.
>
> *De casu diaboli* 25

Introduction

With some preliminaries discussed in Chapter Seven it is time to address the classical statement of the luck problem. First, I do not hope to eradicate the cognitive discomfort in positing an event—the per-willing of A over B (or B over A)—that is not explicable in terms of previous necessitating causes. Certainly the per-willing is caused in the sense that the "motive" power comes from the preceding desire. And the opting for this over that is caused in the sense that it is up to the agent himself—simply by per-willing. As noted in Chapter Seven, even Mele, who insists that agent-causation theories are as damaged by the luck problem as event-causation theories, grants that agent-causalists may have succeeded in defending the thought that the agent-causal agent does have sufficient control over his choice to render the actual choice in the actual world not lucky or chancy. Nevertheless the cognitive discomfort—the *feeling* that all events must have fully explanatory causes—remains. But this discomfort must be born as the price of commitment to created aseity.

The Anselmian insists upon the ability of created agents to make *a se* choices because this is the only way they can engage in self-creation, genuinely contributing something to their own character. In the present chapter I propose that, if aseity is where the central value of freedom and responsibility lies, then proponents of the luck problem have misunderstood the relationship between choices, responsibility, and character. Certainly those advancing the luck problem are free to respond that they do not see commitment to the possibility of self-creation as the chief motivator for defending free will. But the luck problem's *raison d'être* is that it is supposed to undermine libertarianism. I take it that most libertarians hold that the possibility for self-creation, or some not-too-dissimilar phenomenon, constitutes a, if not *the*, reason for developing a libertarian analysis of freedom. If, as I shall suggest, statements of the luck problem assume a relationship between choices, responsibility, and character that is different from what the libertarian—at least the libertarian interested in self-creation—supposes, then, beyond the discomfort already granted, the luck problem does not connect with Anselmian libertarianism.

In the previous chapter I argued that Mele's recent and lengthy statement of the luck problem involving possible worlds does not, in the final analysis, add anything new to the problem as it was explained earlier by Hobart and Hume before him.[1] After Hobart and the debate in *Mind* from the mid 1950s there has been a great deal of discussion of the luck problem, but Hobart's statement remains the recent *locus classicus*, and it will suffice to set up the Anselmian response.[2] The core of Hobart's argument is that it is the self, the individual agent, whom we hold responsible. But if the choice does not come from the agent, that is, it is not traceable to something about the agent before the choice, then there is really nothing there to hold responsible.[3] I noted, in the section on free will rhetoric in Chapter Seven, that the Anselmian picture does not support Hobart's characterization of a choice as something just bursting into being out of nothing. Nevertheless, the Anselmian insists that the *opting* for B *over* A (or A over B) is not produced by anything in the agent's past, or indeed anything except the agent's per-willing B, and it is this *a se* choice which is requisite for responsibility. So if Hobart is right that we should hold you responsible only if your choices are produced by your previous character, then his argument does have traction against Anselmian libertarianism. Hume, whom Hobart cites as inspiration, adds an interesting note which, to my knowledge, has not been adopted and developed by recent philosophers. He points out that acts of choice are ephemeral. They disappear almost as soon as they have come into being. According to Hume, were libertarianism true, then these acts would not arise out of the character of the agent, and so would not leave anything durable behind.[4] Thus, according to Hume and Hobart, libertarianism undermines the very responsibility it had hoped to guarantee.

[1] As I mentioned in Chapter Seven, Augustine brings up a version of the luck problem, but since he does not develop it there seems no need to go back beyond Hume.

[2] For another influential version see van Inwagen's "Rewind" scenario as discussed in Chapter Seven.

[3] R. E. Hobart, "Free Will as Involving Determination and Inconceivable Without It." *Mind* 43 (1934): 1–27.

[4] David Hume, *A Treatise of Human Nature* Part III, Section II.

But the Anselmian takes it that, when it comes to the importance of free choices for responsibility, Hume and Hobart and their more recent followers have the relationship between choices and character exactly backwards. We hold you responsible after you make a choice, not because your choice is symptomatic of a preceding character, but because it creates your subsequent character. The reason that free choices are important is that they make you the sort of person that you are. Choosing is an exercise in self-creation. What we hold you responsible for, and praise or blame you for, is the choice itself and the subsequent character it produces. Your history and character before making an *a se* choice may be very important in determining the degree to which you are responsible, but what you can bear responsibility for are your *a se* choices and their futures, not their history.[5]

The future of the *a se* choice includes doing the chosen deed and then evaluating the deed, and perhaps *choosing* how to think about it. Much of the moral action involved in self-creation occurs after some initial choice has been made, and it is important to get this point on the table. That the Anselmian finds it plausible that your choice creates your character, and that so much of moral import happens *after* you have chosen and acted, may well arise out of a background Christian worldview. In the present chapter I devote some time to noting Christian themes which would support the Anselmian response to the luck problem. I conclude by focusing on one of these themes, forgiveness. Forgiveness poses something of a puzzle for many theories of assigning responsibility, but it fits comfortably within the Anselmian system in which you construct your character through making *a se* choices.

Responsibility and *A Se* Choice

Hobart argues that a responsible choice must flow from a self, and Hume adds the argument that choices, if they are not determined by preceding character, are just fleeting events. Anselm himself notes the transitory nature of actions. "Suppose sin should be . . . an action which someone does, which does not exist except while it is being done, and when it is completed passes away so it no longer exists, . . . As the action passes away so that it no longer exists, the sin would

[5] C. A. Campbell, participating in the *Mind* debate in the 1950s on the indeterminist's side, describes a free choice as a creative act which transcends the preceding character; "Is 'Freewill' a Pseudo-Problem?" *Mind* 60 (1951): 441–65, at 462. Kane mentions a somewhat different, but perhaps related, future orientation when he describes a self-forming choice as an experiment in values, "whose justification lies in the future" (1996: 145).

similarly pass away and no longer exist."[6] (He is speaking of an overt action, not a choice, and his point is that sin does not "evaporate" once the person has done the wicked deed.) And on Anselm's analysis it seems that the choice is even more ephemeral than Hume's description might have suggested in that a choice is the "per-willing" that happens at a fleeting moment and is not even a "thing" with ontological status (see Chapter Four). So aren't the critics right that, if S's opting for B cannot be explained by S's previous character and circumstances, then S should not be held responsible, based on the insubstantial act of opting for B?

Anselm takes it that Hume is wrong to say that the action of choosing leaves nothing that is durable or constant behind it. And Hume and Hobart and their confrères have the moral principle concerning responsibility backwards. If S's choosing B resulted from a God-(and/or nature-) given character, then S could not be blamed for being and doing as God (and/or nature) made him to be and do. What he is blamed for is the character which he *creates* in himself by his choice. For Anselm, the point of being a created free agent is to have a measure of *aseity* and so engage in self-creation and be a metaphysically elevated being. Satan abandons the justice that he could have held onto and wrecks himself. The angels who hold fast make themselves to be better than they were originally made by God.[7] It is right to hold agents responsible for choices and subsequent overt actions, but by the time the agent becomes an object of blame or praise for us, the choice and the action have disappeared into the past. It is plausible, then, to think of the actual locus of the blameworthiness or praiseworthiness *at the time of blame or praise* as the character which the agent creates in himself by making those choices and taking those actions.

Note that many of our common attitudes towards punishment fit well with the Anselmian thesis that we punish you for the character you produce in yourself through your deeds. Contrariwise, a number of (arguably) unwholesome consequences regarding justifications for punishment follow if we suppose that Hume, Hobart, and their camp, have the right analysis. For example, if a wicked act is merely a fleeting symptom of a preceding bad character, and it is this character that deserves the blame, then, if we could know the character before the bad deed, there would be no reason to wait until after the deed to administer the punishment.[8] Or

[6] *De conceptu virginali et de originali peccato (On the Virgin conception and Original Sin)* 4 (my translation; Schmidt II, pp. 144–5, ll.27–5.)

[7] *De casu diaboli (On the Fall of the Devil)* (DCD) 25. St. Augustine found this an appalling thesis and uses it to argue the absurdity of libertarian free will, *City of God* 12.9; *Opus imperfectum contra Julianum* 5.57.

[8] George Sher, *In Praise of Blame* (Oxford: Oxford University Press, 2006): 30–1. Christopher New defends the possibility of "prepunishment"—the legitimacy of punishing you before you have committed the crime; "Time and Punishment." *Analysis* 52 (1992): 35–40. The wrongness of

take, as another example, the standard complaint against consequentialism. Compatibilists, going back at least to the thirteenth century, and including Hume, often adopt the line that punishment is about causing harm to the agent in order to prevent or deter further bad behavior on his part and the part of others.[9] (They often make this move after granting that agents are not the subjects of the very robust form of desert which libertarians hope to preserve.)[10] A standard criticism of this understanding of what justifies punishment notes that, at least possibly, causing harm to innocent agents could achieve this same end and so would be justified.

The Anselmian thesis applied to the justification for punishment avoids these unwholesome possibilities. The wrong-doer, and *only* the wrong-doer, can justifiably be punished, and *only* after he has committed the crime, because the locus of blame is the character he has produced in himself by choosing and doing as he did. If we should say of someone who has freely committed a heinous murder, "He is no longer fit to live among us!" this is presumably because he has committed the heinous murder. Granted, he must have been pretty bad beforehand to be capable of getting to the point where a heinous murder is an option for him. But he was not a murderer until he committed the crime.[11] He can be held responsible for being a murderer if, and only if, he freely chose to commit a murder.

There is an interesting and relevant caveat to the preceding sentence. You can be held responsible for being a murderer only if you freely chose to commit a murder, *and committed it.* We do not punish you for your choices but for your freely chosen deeds. And it is important to note here that, though throughout this work the focus has been on choices, the actual commission of the deed should not be treated as something superfluous to the subsequent formation of your character.[12] It is true that in the Anselmian system, it is the free choice that is most

prepunishment seems to weigh on Hollywood moviemakers in that there have been at least two recent films on the theme: *Minority Report* 2002 and *Captain America: The Winter Soldier* 2014.

[9] Bishop Tempier of Paris condemned compatibilism and this theory of punishment in the Condemnation of 1277, #165.

[10] For example, Daniel Dennett, *Elbow Room* (Cambridge, MA: MIT Press, 1984): 156–65.

[11] For the thesis that it is character caused by bad actions that constitutes the proper ground of punishment see my "Retribution, Forgiveness, and the Character Creation Theory of Punishment." *Social Theory and Practice* 33 (2007b): 75–103.

[12] Of course, in terms of the way the world goes, your doing the deed is enormously important. It may have all sorts of impact on other people and the world in general. But our issues are free choice, responsibility, and self-creation. To the outside world, the impact of the deed could be the same if the mad neurosurgeon were manipulating your limbs, so the brute impact of the deed on the world outside of your choice is not central to our concerns here. What is important for our purposes is the relationship of the deed to your choice and your choice's history and—especially—future.

important. In Anselm's paradigm case of the fall of Satan, Anselm does not even bother to distinguish between Satan's wicked choice and some subsequent deed. (Perhaps this is because Anselm has deliberately and sensibly chosen not to try to fill in any content for possible angelic motivations.) Here is a place where Anselm's idealization of free choice needs to be developed. And in attempting to spell out an Anselmian theory in response to Hobart's version of the luck problem it is well to remember that doing the actual deed after the choice is crucial in several ways.

So, let us move away from Anselm's agent S, and look instead to a human agent, M, for Murderer. A first reason why doing the actual deed is important is that, in real life, there is always the possibility that an agent who has decided upon murder might not go through with it. I am thinking about the sort of scenario in which M was in the torn condition (TC) between murdering and refraining, and decided to murder. Had M been hit and killed by a truck an instant after making that decision he might, in the eyes of God, be guilty of having chosen to murder (there is a qualification below). But suppose that, upon arriving at the point where he is aiming the gun at his victim, M finds something new in the situation to consider; perhaps the pleas of the victim, or a photo indicating that the victim has young children, or just the sudden realization that he, M, is actually at the point of destroying a human life. The new situation might throw M into a new TC and perhaps he will opt to change his mind and not commit the murder. One can envision a case in which M chooses to murder with such a strong commitment that nothing can shake it. He cannot find himself in a new TC with respect to the murder. In that kind of case the sort of new situation I have described cannot arise. But only God can know that some instance of M's choosing to murder involves such a strong commitment. For the rest of us, even for M himself, it may take M's actually murdering for us to know that M really *chose* to murder.

This is not just an epistemic point. Given the suggestion of the "new situation" above, until he has actually gone through with the murder, it may not be the case that M has truly per-willed to commit murder. Up to that point he may be simply willing (or per-willing) the preliminaries to committing a murder. Someone who chooses in an irrevocable way to commit a murder may, from the God's-eye point of view, be a murderer without ever committing a murder. But absent that unshakeable choice, you become a murderer only by choosing to murder all the way through to the murdering. This is presumably one of the reasons that we punish attempts less than actual murders. At least in some sorts of cases, for example those in which you were stopped before you could carry out the actions which would constitute the murder, we cannot know whether or not you would have gone through with it. More importantly from the perspective of self-creation, in that you *didn't* go through with murder, you may not have per-willed to murder in such a way as to form your character as a murderer.[13] The importance of actually doing the deed underscores the point that, *pace* Hume and Hobart, we are not holding you responsible because your choice and deed flowed from your character before your deed. We are holding you responsible because you did the deed you had freely chosen to do.

There is a second reason that actually doing the deed is crucial in the process of self-creation, and that has to do with how you think about your deed after it is done.[14] It seems to me, and here I am working simply from introspection, that a very significant part of the process of character-building takes place after the fact of an overt action.[15] Note, again, the importance of the *future* of the choice and deed. M, if he succeeds in murdering, must now confront his past action. Theoretically it is conceivable that M's choice to commit the murder shapes his character as a murderer such that it is fixed. But more likely he finds himself in a new TC, but now the options are constituted by the various attitudes he could adopt towards his murder. For example, he could choose to embrace it and glory in the evil. Or he could embrace it as somehow justified; he could try to convince himself that the victim deserved it, etc. Or he could embrace it as something he couldn't help but do; he could try to convince himself that Hobart and Hume are correct and he was driven to murder by his character which was formed by, for example, the poor treatment he received as a child. Or he could reject the murder and repent. Or he might adopt some other attitude. The point is that in terms of self-creation a great deal of the tearing down and building up is likely to take place as one ruminates on one's past deeds and—at least in the Anselmian system—freely chooses to adopt certain attitudes towards them. This is not a backwards-looking process. One commits to trying to be a certain sort of person in the future based on how one views one's past actions.

Evaluating one's past actions is another venue in which attempts will function differently from successes. Suppose M is stopped before he can commit the

[13] Rogers (2007b): 92–4.

[14] Self-evaluation is often a difficult process and one may fail, for a variety of reasons, to assess oneself, and one's character, choices, and deeds correctly. Nevertheless it is an essential part of the task of self-creation. A discussion of the topic lies outside the scope of this work, but see Anselm's *Prayers and Meditations* for interesting examples of introspection and evaluation of one's thoughts and deeds. The modern reader of this work is likely to conclude that Anselm was too hard on himself. Perhaps that is one of the burdens of being a saint.

[15] Apparently there is some evidence to the effect that when people in general assess degrees of responsibility they focus on the agent's identifying with his action. Tamler Sommers, "Experimental Philosophy and Free Will." *Philosophy Compass* 5 (2010): 199–212, at 207–8.

murder. He may grind his teeth in anger at being stopped, he may thank God that he was unable to go through with it, etc. If the aftermath of an overt deed is very significant for self-creation, then whether or not one succeeded or merely attempted (assuming one is aware that one has not succeeded) must play a role in the self-assessment. If I am right that a lot of the moral action in character-building takes place after the agent does the deed, then this supports the Anselmian claim that what you are responsible for is not the character—or based on the character—you had *before* you freely chose. Rather you are responsible for your choice, your deed, and the character you formed in yourself by the choice, the deed, *and* the subsequent choices you make about how to evaluate your deed.

The central motivation for Anselm's construction of his libertarian theory is to elaborate a view which allows for created agents to self-create. And self-creation requires aseity. But aseity for created agents entails confronting open options, with nothing in their past to determine them to one over the other. And that opens the door to the luck problem. Anselm himself grants that this theory entails an element of "mystery." Why did Satan choose as he did? Only because he chose, "if such a thing can be said." But, as I argued above, there are good reasons to believe that Hume's and Hobart's view is mistaken. Our ordinary assumptions about punishment conflict with the thought that it must be the agent's character prior to a choice which constitutes the locus of responsibility. If the Anselmian is correct that what we are actually holding you responsible for is your choice and the character it forms, then, although the cognitive discomfort remains, the effectiveness of the standard version of the luck problem is undermined.

Background Beliefs

It seems to me that this debate between the Anselmian on one side and Hume and Hobart on the other may well provide another instance where fundamental disagreements in intuitions grow out of an underlying difference in views about the world and the situation of human beings in it. There are a host of background beliefs that, in one way and another, bolster Anselm's commitment to the claim that what you are responsible for in assessing a particular choice is not rooted in your pre-choice history, but rather in your choice and the character it produces. (Though you may be responsible for the pre-choice history insofar as that is a product of earlier *a se* choices.) And it may be that, for some working on the free will question, either sharing or rejecting these (or relevantly similar) background beliefs plays some role in driving their intuitions. I do not say that canvassing these beliefs is likely to make Anselmian libertarianism more persuasive for those who reject his worldview. Probably the contrary is the case. But, as I explained in the introductory chapter, spelling out background beliefs seems to me a useful exercise in the campaign to encourage philosophers working on free will to state the assumptions in which their intuitions may be rooted. Think how much time it can save in the future if it turns out that no amount of tinkering with examples involving goddesses and mad neurosurgeons will change any intuitions because the real disagreement is over whether we live in a theist universe or a universe only of atoms and the void (or pick some other popular universe)! And, should it become clear that disagreements in intuition are *not* rooted in differences in basic worldview, well, that would be good to know, too.

So let us look at some of the background beliefs that may help give rise to Anselm's taking it for granted that free choices produce character, and that it is the choice and the *subsequent* character that constitute the locus of responsibility. Note first that, as a classical theist, Anselm believes that God is eternal and immutable. That means that God just doesn't have a history. Nonetheless, as Anselm sees it, God is radically free in that all of his acts of will (which are in fact only one act which is identical with His nature) are absolutely *a Se.*[16] And Anselm makes a point of defining "free will" so that the same definition is applicable both to God and human beings.[17] Thus, the thought that freedom must arise out of a certain sort of history will not occur to Anselm.

Further, as I noted in the Introduction to this book, Anselm answers the question, "Why do we want free will?" in a very different way from many contemporary philosophers. Many contemporary philosophers are most concerned to defend a theory of free will which will allow us to justify our usual beliefs and practices regarding social and legal praise and blame and reward and punishment. If this is the central concern, then there is a practical problem with focusing, as the Anselmian does, on the choice and the subsequent character. We—society, those around the agent—cannot observe the agent's choice nor his subsequent character. If our ultimate concern is how *we* should treat you, then locating responsibility in these two "invisible" phenomena seems pointless, and a theory of free will which entails this looks unhelpful. Taking the observable overt deed to be a symptom of your choice-preceding character, as Hume and Hobart seem to do, might be thought to fit better with a discussion motivated by questions about how human third parties should deal with agents.

The Anselmian, of course, is interested in these questions, but more important than how we treat you is what sort of person you actually *are*. But isn't this just ultimately invisible to us? Anselm, and those who share his worldview, do not

[16] Rogers (2008): 185–92. [17] *De libertati arbitrii* 1.

take *our* opinion of the agent to be the most important opinion. How the agent sees himself is more important. Presumably he has an insider's perspective, and careful introspection is a key aspect of the job of building one's own character.[18] And even *more* important, there is a third party who is privy to the agent's choices and character, and that is God. So the agent's choices and subsequent character are quite visible to the Observer who matters most. Anselm's analysis of free will has theism in its background worldview while Hume's (and many of his followers') does not, and perhaps this colors intuitions about the right way to understand the direction of the causal arrow regarding choice, character, and responsibility.

Certainly core Christian doctrines require embracing a theory more like Anselm's than Hume's. Anselm's paradigmatic instance of choice is the fall of Satan. The Christian might feel that angelic business need not be part of his philosophical investigations. And some modern Christians seem content to abandon the story of the human fall, the doctrine of original sin, and the thought that things might go very badly for someone posthumously and forever. But traditional Christians will not lightly throw away these core Christian doctrines, and even for those who absolutely reject them, these themes—perhaps just as cultural or literary memes—may evoke some echo in their intellectual endeavors. The thesis of the fall is that God made human beings good and human beings, by their bad choices, plunged themselves into the dreadful condition in which we find ourselves. This narrative is impossible to square with Hume's and Hobart's view where your blameworthy character must precede and determine your choice. And then there's hell. The Catechism of the Catholic Church (section 1033) defines "hell" as the "state of definitive *self-exclusion* [my italics] from communion with God and the blessed." The Anselmian tradition holds that you make your way to hell by what you choose and do.[19] Again, this is impossible to square with Hume's and Hobart's understanding.

Forgiveness

And finally, there is a further, much happier, side to traditional Christianity which may also underlie the thought that it is the future of the choice that counts,

[18] Anselm's Prayers and Meditations make this very clear.

[19] The doctrine of original sin has it that you are inevitably bound for hell unless God offers you grace. Anselm himself followed Augustine in saying that God does not offer grace to everyone. However, there is no harm done to the Anselmian system if—finding the doctrine of original sin of significant explanatory value—one holds that God does offer grace to everyone. Rogers (2008): 141–5.

not its history. One of the major points of the Christian story is repentance and forgiveness—vis-à-vis God and vis-à-vis our fellow human beings. Whatever evil you have done, and whatever defects of character may have helped to bring you to the point where evil could be an option for you, the guilt can be undone. In one sense you are always responsible for the choice and the deed. You did it, after all. But, on the Christian accounting, you can cease to be blameworthy. If you come to deeply regret and utterly hate that wicked choice and deed, then you can uproot its impact from your character and be forgiven. On the Anselmian understanding, a free act of repentance constitutes a new choice you make after your earlier choice for evil. Instead of embracing your evil choice, you choose to reject it.

Note first that it is the choice and deed which are the accepted and appropriate objects of repentance. You might repent of a way of life, but that must be sorrow over a series of choices. Repentance makes sense only if it is regret over what was, in some way, up to you. I take it that even compatibilists will agree with this point about reactive attitudes. If you believe that you are subject to a vice which preexists, and hence could not have arisen out of, your choices, then you might be unhappy about it or wish you didn't have it, but it would be incoherent for you to *repent* of it. Relative to the mere possession of this vice, you haven't *done anything* to be sorry for. Hume seems to be suggesting that a choice has an effect on your subsequent character only when the choice is determined by your preceding character. (His brief remarks do not make it perfectly clear, and I leave it to Hume scholars to assess how his remarks on the locus of responsibility are to be squared with his "bundle" theory of personal identity.) But then it seems that the clear-sighted agent ought not to repent of bad choices and subsequent bad character. The subsequent character is produced by the choice and the choice is produced by the preceding character and the preceding character wasn't anything that the agent did. You cannot reasonably repent of something you did not bring about. That, at any rate, is how the Anselmian sees it.[20]

The compatibilist may respond that the Anselmian is begging the question. He may hold that the attitude of repentance can indeed be appropriate concerning a character and choices necessitated by the pre-choice condition of the agent. Here we are back to the original and fundamental disagreement between the

[20] Derk Pereboom notes a similar point, but supposes that the negative feelings consistent with believing in hard determinism might suffice for our going about our business; *Living Without Free Will* (Cambridge: Cambridge University Press, 2001): 204–7. The Anselmian responds that it is one thing to be sorry for what you yourself did and quite another to regret that the universe goes in such a way that it produced your doing some bad thing. The latter cannot stand in for the former.

Anselmian and the compatibilist. In Chapter One I argued the incompatibilist side, casting this disagreement in Anselm's own terms where, if there is a necessitating cause for the choices of human agents, it is God. On this question of repentance we can return to Anselm's own vivid example and ask: Wouldn't it be absurd for Satan to sincerely and deeply regret choosing B if he correctly believes that his choice for B was caused by God? If he correctly believes that God made him do it, then he believes that his choice for B was all for the best somehow. But genuine repentance entails a wish that the choice had (really and absolutely) not been made. Perhaps a compatibilist Satan can reasonably engage in some repentance-like behavior, but he cannot coherently repent. *Mutatis mutandis*, the naturalist compatibilist agent must recognize that his choice is the necessary product of the past and the laws of nature. To *truly* wish it away is (to borrow William James' point)[21] to wish away the whole universe. On compatibilism, "I wish I had not chosen that" entails, "I wish the actual universe had never existed." The naturalist compatibilist may have all sorts of regret-like feelings, and perhaps such feelings do not signal any irrationality on his part, but *robust* regret—a sincere wish that the choice *actually* not have occurred—is just not a coherent option for the compatibilist. But repentance is an ineradicable aspect of the Christian experience, and it may be that this colors the Anselmian intuition that what you are responsible for is your choice and subsequent character.[22]

And then there is the matter of forgiveness. What I mean by forgiveness here is recognition—perhaps by a third party, but also perhaps by the agent himself—that though the agent freely did the blameworthy deed, and may have produced in himself a subsequent blameworthy character, now that he has repented he is no longer blameworthy. His repentance helps to "wash away" the effect of the choice. He is still responsible in the sense that he did it, but he is not responsible in the sense of being guilty of it or blameworthy for it. On the Anselmian self-creation model, the agent is able to tear down some of the edifice which is his own character and rebuild it along more wholesome lines.

The Christian story is driven by the claim that God forgives those who accept His gratuitous help. And a main pillar of Christian ethics is that we are to forgive one another. It crops up in our legal system when the case is made that, due to apparently sincere repentance, evidenced by good behavior in the time after the commission of a crime, an offender does not deserve as much punishment as

[21] William James, "The Dilemma of Determinism." *Unitarian Review* 22 (1884): 193–224.

[22] How Augustine and Aquinas and Calvin might square this point with their compatibilism is a problem I leave for others. I do not see how it can be done.

might have originally seemed appropriate. For example, an offender may be released early due to good behavior in prison. This may be in part because he is no longer judged to be a danger, but it may also be due to the thought that he no longer deserves punishment. He has ceased to be blameworthy. And sometimes an offender who is not caught until years after his crime receives a lesser sentence than he would have had he been caught immediately because he has lived the life of a decent citizen in the meanwhile. Again, he may be judged to be less of a danger, but the thought that he simply does not deserve the more severe punishment seems to play a role.

It is worth noting that the claim that it is appropriate to forgive the repentant wrong-doer poses a puzzle for many theories of punishment. Consequentialists are likely to argue that the beneficial effect of punishment would be vitiated if offenders could get off the hook by repenting, and retributivists tend to hold that if you did the deed you must pay the piper. Period.[23] Forgiveness is such a puzzle that some contemporary philosophers, recognizing the reasonableness of the intuition that someone who has undergone a radical change of heart deserves less punishment than they originally deserved, have argued that the changed person is, in some more-than-metaphorical sense, a genuinely *new* person. Alwynne Smart suggests that in this case, the "real offender no longer 'exists' or fully exists."[24]

On the Hume–Hobart analysis, where the locus of responsibility is the choice and its pre-history in the preceding character of the offender, if one felt compelled to defend the value of forgiving the repentant offender, perhaps one would have to make Smart's move and say that the offender no longer exists. But this is not a very plausible approach. Surely, if you borrowed $50 from the evil-doer, you still ought to pay him back if he repents of his evil-doing. It seems absurd to hold that he is a different person. Indeed, if he really is not the same person that did the evil deed, then what is he repentant for? Excepting the silly public apologies for others' wrong-doing that politicians go in for, we do not repent of the evil deeds of others.[25] But if the locus of responsibility is the choice and the subsequent character which the choice creates in the agent, then genuine repentance—freely chosen by the agent—constitutes the agent re-working his character so that the bad effects of the wicked choice are undone, and forgiveness is the appropriate response.

[23] Rogers (2007b): 94–5.

[24] Alwynne Smart, "Mercy." *Philosophy* 43 (1968): 345–59.

[25] Perhaps an exception is the parent apologizing for the deeds of the child. But, presumably, this would be appropriate because the parent feels responsible for contributing to the character and behavior of the child.

There may be good consequentialist reasons for those tasked with protecting society to take a cautious approach to forgiveness. To note just one major problem, repentance can be feigned. If our main motivation in discussing free will were an interest in grounding consequentialist justifications for legal punishment, then issues revolving around repentance and forgiveness would likely play little role. But within the Christian worldview the most fundamental question regarding praise and blame has to do with what agents actually deserve based on what sort of people they are at the very time that praise and blame become an issue—that is *after* a choice is made and a deed is done. And issues of repentance and forgiveness loom large in thinking about what the agent actually deserves. And so again, the background perspective motivating one's interest in the free will question may play a significant role in grounding one's intuitions.

The Anselmian concludes that the case made by Hume and Hobart that responsibility requires that the choice be determined by the preceding character just has the relationship between choice and responsibility backwards. The agent is responsible for the subsequent character he creates in himself through his free choices. True, the element of aseity generates some cognitive discomfort. Though S's choice for B is both caused and explicable, in that we can point to S's desire for B and his per-willing B, the Anselmian must allow that there is nothing in S's past which makes it the case that he chooses B at t2 rather than A. And perhaps it is human nature to feel disquiet at the thought of an event which is not causally necessitated. But this disquiet does not uproot the basic agent-causal claim that if S chooses B, then choosing B is something that S *does*, it is not a bit of luck that happened to him. The disquiet does not constitute a reason to abandon libertarianism.

The luck problem was supposed to intuitively undermine the responsibility of the libertarian agent. I have argued that describing a libertarian choice as a chancy sort of event begs the intuitive question against the libertarian. Mele's casting the luck problem in terms of possible worlds does not seem to add anything to the original problem once we try to elaborate on the possible worlds locution. And a plausible argument can be made that the original problem as set out by Hobart and Hume just locates the point(s) of responsibility at the wrong place in the timeline of a choice. Admittedly, the back and forth centering on the luck problem is a battle of intuitions. I hope at a minimum to have clarified some of the background beliefs and commitments which may drive the intuitions on the Anselmian's side of the issue. Perhaps the proponents of the luck problem can do the same, and it will become clear whether or not there is a brute difference in intuitions about free choice. Perhaps the differing intuitions are based in disagreements about the fundamental nature of the world.

9

The Tracing Problem

> [The good] angels are not to be praised for their justice due to the fact that they were able to sin, but rather due to the fact that, in a way, they have it from themselves that they are [now] unable to sin; in this they are, to some extent, similar to God, who has whatever He has from Himself (*a se*).
>
> *Cur deus homo* 2.10

Introduction

The quotation above is the one with which I began this book. It sums up what Anselm and the Anselmian take to be the reason we want freedom—the freedom which requires *a se* choice; without it we cannot be the sort of metaphysically valuable agents who can contribute to their own creation. This claim allowed me to argue, in the previous chapter, that Hume, Hobart, and their followers who raise the luck problem, have the relationship between choices, character, and responsibility backwards. Your responsibility does not require that your previous character necessitates your choices. Rather, you are responsible for the subsequent character which you produce in yourself by your choices. Thus the Anselmian holds that you can be responsible for character-determined choices, assuming the character was formed through the sort of *a se* choices which ground responsibility. Responsibility is "traceable" through self-formed character to the *a se* choices which produced that character. Anselm seems to take this tracing thesis for granted, and it is central to the Anselmian view that what human freedom is *for* is self-creation.

But a dilemma arises here which, if left unresolved, would undermine the basic Anselmian thesis about self-creation. It seems that you cannot be responsible for something done in ignorance. But how likely is it that you (or people in general) are aware that choices produce character? In the present chapter I explain the dilemma and then attempt to solve it. I dismiss the possible "solution" which involves prematurely surrendering and simply allowing that perhaps most agents cannot be held responsible for character-determined choices. Next I look at

Aristotle's solution which is that those who don't know that choices form character *ought* to know. In that Anselm assumes the tracing thesis, but does not make a point of defending it, or even of spelling it out at much length, he may have simply inherited it from earlier thinkers, and so may be an Aristotelian on this question—though at some remove. Indeed, as applied to most adults Aristotle's response seems plausible. However I believe that there are some counter-examples. In that it is better to include people within the fold of moral agency, I attempt a third possible solution: We can make a principled distinction between ignorance which is exculpatory and ignorance which does not undermine responsibility. I will argue—not based on Anselm's own work, but using aspects of his system as I have tried to spell it out—that ignorance concerning the fact (at least the Anselmian takes it to be a fact) that choices produce character is not the sort of ignorance that would undermine responsibility for one's choice-produced character. If the argument is successful, then the tracing thesis, so central to the Anselmian theory, stands firm.

The Dilemma of Ignorance and Tracing

As set out in Chapters Four and Eight, a central component of the Anselmian system is that what you bear responsibility for is your *a se* choices, the character which they produce in you, and the choices which are determined by your self-created character. As I noted, this is a point which Anselm himself takes more or less for granted. In the idealized instance of choice which Anselm chooses as his paradigm, the good angels forever set their character for good by clinging to the justice that God has given them, and Satan forever sets his character for ill, by rejecting this justice. Even though they can no longer find themselves in the torn condition, since the good must inevitably choose the good and the bad choose the bad, the angels are nevertheless free, responsible, and deserving of praise or blame. Their subsequent freedom is grounded in the fact that it was through their *a se* choices that they established their characters. Anselm's example of self-forming choice is extreme, but the thought that character is formed by choice is common, going back, as noted, at least to Aristotle. Common, too, is the entailment that you may be responsible for character-determined choices if you are responsible for your character—the "tracing thesis."

And it is generally acknowledged, again going back at least as far as Aristotle, that responsibility requires some understanding of what you are doing and what the likely consequences of your action will be. In Anselm's paradigm instance of choice, Satan recognizes that he could conform to the divine will and that, in choosing what he chooses, he violates a divine command. Were he genuinely and innocently unaware of this, he could not face the morally significant options that set up the possibility for *a se* choice. Call the principle that you must (to some extent) understand what you are choosing the "epistemic requirement."

A dilemma arises when we couple this epistemic requirement with the tracing thesis. Granting that choices form character, it is unlikely that you knew, in making many of your earlier, character-forming choices, what the consequences of those choices would be in terms of character formation and future, character-determined choices. But if you didn't grasp what the consequences of your action would be, it seems that you should not be held responsible for forming your character and for consequent, character-determined choices. Thus ignorance poses a problem for the tracing thesis.[1] I will argue that while this ignorance may affect the degree to which an agent is responsible for character and character-determined choices, it is nonetheless consistent with basic responsibility.[2]

A great deal might be said concerning the epistemic requirement, but present purposes do not demand a catalogue of the necessary and sufficient epistemic conditions. A rough and intuitive discussion will suffice. (Later in the chapter I will attempt to distinguish a category of rational belief that is *not* required for responsibility). Aristotle, as usual, has some helpful things to say concerning the relationship between ignorance and responsibility, though I will suggest that he does not provide a fully adequate solution to the tracing problem.[3] He asks us to consider the case of drunkenness. Say that you are drunk and hence unable to

[1] See Manuel Vargas, "The Trouble with Tracing." In *Midwest Studies in Philosophy 26: Free Will and Moral Responsibility*, edited by Peter A. French, Howard K. Wettstein, John Martin Fischer, Guest Editor (Oxford: Blackwell Publishing, 2005): 269–91. Vargas points to trouble with tracing regarding responsibility for specific, character-determined choices. I address this concern briefly, but concentrate on responsibility for the character formation itself. For work that addresses Vargas' approach at length see John Fischer and Neal Tognazzi, "The Truth about Tracing." *Noûs* 43 (2009): 531–56; Kevin Timpe, "Tracing and the Epistemic Condition on Moral Responsibility." *Modern Schoolman* 88 (2011): 5–28.

[2] Peter van Inwagen casts doubt on the thesis that certain mitigating factors result in degrees of responsibility; "Genes, Statistics, and Desert." In *Genetics and Criminal Behavior* edited by David Wasserman and Robert Wachbroit (Cambridge: Cambridge University Press, 2001): 225–42. True, an agent is either responsible or he is not. But when an agent *is* responsible it seems appropriate to hold him more or less responsible in the sense that his blameworthiness or praiseworthiness is subject to all sorts of relevant facts, such as age and mental acuity.

[3] *Nicomachean Ethics* III.5. For scholarly discussion of this text see David Bostock, *Aristotle's Ethics* (Oxford: Oxford University Press, 2000): 117–18; Sarah Broadie, *Ethics with Aristotle* (Oxford: Oxford University Press, 1991): 164–74; T. H. Irwin, "Reason and Responsibility in Aristotle." In *Essays on Aristotle's Ethics* edited by Amelie Oksenberg Rorty (Berkeley, CA: University of California Press, 1980): 117–55; Michael Pakaluk, *Aristotle's Nicomachean Ethics* (Cambridge: Cambridge University Press, 2005): 143–9; Jean Roberts, "Aristotle on Responsibility for Action and Character," *Ancient Philosophy* 9 (1989): 23–36.

make rational decisions and you do some bad deed—to take a contemporary example, you get drunk, drive your car, and accidentally hit and kill a pedestrian due to your drunkenness. Aristotle would have it that, even if you were past the point of rational decision-making when you chose to drive, you are responsible for the homicide, since you are responsible for getting drunk. True, while you were getting drunk you didn't anticipate that this particular pedestrian would be in the street at this particular intersection. Nonetheless, you understood in general that situations like this could arise, and still you drank. You are not as responsible for the death as you would be if you had deliberately set out to kill the pedestrian, but you are responsible.[4] That all seems intuitively plausible. If you are responsible for getting drunk, you are responsible—at least to some degree—for whatever bad consequences your drunken state occasions. Analogously, the Anselmian holds that if you construct your character—for ill or for good—through responsibly made choices, you are responsible for your character and for the character-determined choices to which your character leads you, even if, at the time you were forming it, the future situations in which those choices would be made were unknown to you.

The critic may question this approach and argue that, with drunkenness and character formation, ignorance of what the future holds *does* constitute an excuse, in which case we cannot trace your responsibility for a present choice or act back to the responsibly made choices which resulted in the drunkenness, or the character, that produced it. The Anselmian responds that this is unreasonable. If the proposed exculpating factor is ignorance of what *particular* state of affairs will obtain in the future, then we would have to disallow responsibility in all sorts of cases where we would normally assign it. We are almost always deeply ignorant of what the future holds. Take an example from a life project similar to developing character—or perhaps it is really one aspect of developing character—that is, getting an education. The goal of much of education is to get students to internalize methods, concepts, and principles to the extent that they can apply them in new situations, including situations they may not have previously envisioned.

Suppose a raised bike path collapses due to sloppy engineering. We discover that the engineer in question goes about his job in a generally sloppy way relative to the standards of engineering. We further find that he learned sloppy

[4] Aristotle himself says that penalties are doubled in the case of drunkenness. He is focusing on ignorance of the law, and so perhaps the situation he is envisioning is unlike the homicide case. In the homicide case contemporary law and common intuition would note the mitigating factor that you did not intend to kill anyone. Still, we do consider drunkenness worthy of blame, and we punish driving drunk, even if you do not break any other law.

engineering habits due to his own, responsibly chosen, cutting corners, cheating, and general laxness as he studied engineering in college. He practiced his engineering vices so well that now he cannot help but be sloppy in his work. Can he argue that he is not responsible for the collapse of the bike path because his sloppiness in this particular case was determined by his habitual sloppiness, and he did not foresee having to build a raised bike path as he was developing his sloppy habits? Surely not! Of course you cannot foresee all the specific situations in which you will find yourself. The problem is that the engineer has sloppy habits. He formed those habits responsibly, and so he is at fault.

If the engineering student were non-culpably ignorant of the fact that he would be asked to build *anything* in the future, then perhaps he is not responsible for forming his poor engineering habits. Say our student was forced to study engineering by his parents, assumed he would never get a job as an engineer, developed the bad habits on this assumption, and then, unfortunately and despite his best efforts, found himself employed in building things. In that case, perhaps, he is not responsible due to his lax engineering habits. (What he is blameworthy for is accepting, and continuing in, a job for which he should know he does not have the required skills.) But, except in bizarre cases like this, the engineer is responsible for the collapse of the bike path, even though he did not foresee, as he developed his bad engineering habits, that he might be called upon to build this bike path.

In education in particular, but also regarding life in general, we spend much of our efforts preparing, one way and another, for an uncertain future. Predicating responsibility on detailed foresight makes an impossible demand. Thus, invoking our ignorance of particular future situations to deny responsibility for character-determined choices is counterintuitive. The Anselmian's Aristotelian intuition about tracing is plausible, and if the tracing problem were focused only on our ignorance of particular future situations in which we might have to make choices, it might not seem too much of a problem.[5]

But getting an education and developing your character do come apart in an interesting way relevant to the epistemic requirement as a problem for tracing. If you are engaged in getting an education in engineering then, barring the most outré imaginings, you cannot be ignorant of the fact that you are getting an education in engineering.[6] But suppose it is not universally obvious to agents

[5] Fischer and Tognazzi (2009: 537–8) mention this point, but they do not address the question of ignorance concerning the fact that you are forming your character by making choices.

[6] The ingenious philosopher may suggest exotic cases. For example, suppose, unknown to you, someone slips you a pill which produces the same results in you as many years of study in

engaged in character-forming choices that they are thereby building their characters. What if their ignorance is not about particular future situations they may face, but about the very fact that they are forming a certain character by their choices? Here the example of drunkenness as it was set out above, to motivate the Aristotelian intuition, is not analogous. In the example it is supposed that the agent knew that drinking would, or at least might, result in his being drunk. On the suggestion that an agent might be ignorant of the character-forming effects of his choices, a closer analogy, still using the example of drunkenness, looks to be the situation in which you did not know you were getting drunk. (I will ultimately argue that this is *not* an analogous case). A first response might be that you ought to have known! But take this example: Intending to stay functionally sober, you stop drinking after one, very small, drink. However, unknown to you, an acquaintance has slipped a pill into your drink—a pill which magnifies the effects of alcohol twenty-fold. Now you are drunk. And you did in fact *choose* to take the drink that made you drunk. But you could not have anticipated that you would get drunk. Suppose that the one, doctored drink sends you past the point of rational decision-making, you get behind the wheel, and hit the pedestrian. The law, for consequentialist purposes, may have to treat you as if you were responsible, but intuitively it seems clear that you are not. You were ignorant that drinking the one drink would get you drunk, so you are not responsible for being drunk. Analogously, if we do not see that our choices produce the virtues and vices that will shape our future lives, it seems we should not be held responsible for our characters. And since it was responsibility for our characters that constituted the ground for our responsibility for character-determined choices, we should not be held responsible for our character-determined choices. If that is the case then the project of responsible self-creation must fail, at least for those who have not *deliberately* set out to engage in the project.

Aristotle's Solution

There are three moves the Anselmian defender of the tracing thesis could make. The first involves premature surrender. The second is Aristotle's solution, which serves for many, or even most, cases, but perhaps not all. The third is a new suggestion involving what I take to be a plausible limitation on the epistemic requirement. It is a suggestion that accords well with Anselm's understanding of a choice as a "thin" event, as spelled out in Chapter Four.

engineering. Just by swallowing the pill you get an education unawares. Such frivolous counter-examples do not undermine the distinction I am drawing.

The first possible move is to admit that character is often formed in ignorance, allow that this ignorance is exculpatory, and admit that agents are rarely responsible for character-determined choices. This move is unattractive in that, at this stage, it is too early to abandon the well-established and intuitively appealing view that we *are* often responsible for our character-determined choices. In the Introduction to this work I proposed the methodological principle that inclusiveness is an advantage in a theory of free will. Being a responsible agent is a good thing, even if it does sometimes entail being the subject of blame and punishment. For one thing, being responsible is requisite for being *praiseworthy*. Moreover, at least in some traditions, including and especially the Christian tradition within which Anselmian libertarianism was born, responsibility is a key aspect of human dignity. In this tradition, to deny that an agent is responsible for many of his choices is to demean him. Assuming that we do not want to demean our fellows without good cause, we should not settle for this first response unless other possibilities fail to satisfy.

A second approach holds that being unaware that choices form character is *culpable* ignorance. Unlike with the spiked drink, if people do not understand that choices cause character, it is their own fault.[7] Aristotle seems convinced that any reasonable person recognizes that he shapes his character by his actions. Assuming choices are actions of a sort, and that choices and overt actions are usually closely related, we may apply the Aristotelian approach to both overt actions and choices.[8] Aristotle writes, "Now not to know that it is from the exercise of activities on particular objects that states of character are produced is the mark of a thoroughly senseless person."[9] David Bostock, discussing this passage, asks what to say about people who really do not know that actions produce dispositions. "In these cases the disposition is perhaps acquired 'in ignorance'. But may we not say that such people *ought* not to be so ignorant? What they should do is read what Aristotle has said in book II [of the *Nicomachean Ethics*]."[10]

Now, certainly, everyone should read Aristotle. And it is not absurd to claim that a reasonable person ought to understand that choices and overt actions

[7] This is Robert Kane's preferred solution to the tracing problem in "Three Freedoms, Free Will and Self-formation." In *Essays on Free Will and Moral Responsibility* edited by Nick Takakis and Daniel Cohen (Newcastle on Tyne: Cambridge Scholars Publishing, 2008): 142–62.

[8] One might make choices concerning internal states, for example, "From now on, I will try not to feel envious!" And presumably here, too, practice helps form virtues and vices.

[9] *Nicomachean Ethics* III.5, 1114a. I follow the W. D. Ross/J. O. Urmson translation, in *The Complete Works of Aristotle* Vol. 2, edited by Jonathan Barnes (Princeton, NJ: Princeton University Press, 1984): 1759.

[10] Bostock (2000): 117–18. Irwin (1980: 141) also seems sympathetic to Aristotle's claim.

produce character. In one sense, of course, it is a truism that one who does a certain sort of action consistently is a that-sort-of-action-doing kind of person. But the Anselmian defender of the tracing thesis must intend more than this tautology given his analysis of character and responsibility where virtues and vices are robust traits with explanatory powers. In saying that the character determines the choice we are distinguishing between the character and the choice that is determined, and we are granting to the character a significant efficacy. Moreover, in the Introduction I noted that the reason the Anselmian wants free will has more to do with what kind of person you are than with the discreet actions you perform. So the Anselmian must hold that choices and actions produce a robust character beyond the tautology about consistent actions. The question is: Should we hold that even those unfortunates who were never exposed to Aristotle *ought to* grasp that choices and overt actions produce virtues and vices? They are blameworthy for their ignorance and hence can be held responsible for character-determined choices?[11]

Anselm's analysis of morally significant choice presupposes that the rational agent is self-reflective. As I noted in Chapter Three, the torn condition that must precede the free choice involves the competing desires for justice and some mere benefit inconsistent with justice. And justice, in Anselm's schema, is a second-order desire about first-order desires for the appropriate benefits. A glance at Anselm's *Prayers and Meditations* shows that he takes self-examination to be a standard project in the Christian life. With this emphasis on self-reflection, Anselm himself might well suppose that moral agents ought to recognize that choices form character and that they are blameworthy if they do not see this.[12] And within the Christian tradition even children are taught to adopt a self-reflective attitude. Even the Lord's Prayer, ubiquitous in Christian worship, asks God to "lead us not into temptation." That is a prayer about what sort of desires we hope to have, and so it involves awareness of an inner life. So even setting aside commitment to Anselm's hierarchical schema for free choice, the vast majority of Christians young and old—and perhaps even those who have just grown up in, or are conversant with, the Christian tradition—are likely to embrace the related thoughts that you ought to want to be a certain sort of

[11] This suggests an interesting question: Suppose someone is blameworthy for their ignorance concerning character-formation, but suppose that the character formed and the subsequent character-determined acts are praiseworthy? There is nothing contradictory here, but it seems a bit strange.

[12] See my "Anselm on Self-reflection in Theory and Practice." *The Saint Anselm Journal* 9 (2013) [Online].

person, that you ought to want to have a certain sort of interior life, and that your choices and actions help to construct your character.

Moreover, the process of education suggests that the Aristotelian thesis about character formation is readily available to most people. For example, the parent who wants to start the child on the path to becoming compassionate will say, "Just think of how you would feel if someone did that to you!" That is, the child is told to do something, to try to enter imaginatively into the feelings of the other. The hope is that practice at noticing that there are other people in the world, and that their feelings are to be considered, will lead to the child's eventually coming to take account of the other's feelings on his own and habitually. If you want to teach a child to be generous, the initial lessons in sharing tend to involve the use of force. First you voice the thought that, "It would be so nice to share that with your friend." Then you wrest a bit of what is to be shared out of the grip of the puzzled and indignant child, hand it to the little friend, and—perhaps most importantly in light of the lesson to be taught—praise the child who "shared" for his generosity. One hopes that after a few (dozen? hundred?) repetitions it will occur to the child to share. Those who are acquainted with Aristotle will recognize these examples as instances of the method which he proposes for learning to behave well; what it takes is practice in choosing and doing the appropriate actions under the guidance of someone who already understands. But parents all over the globe have always adopted this method. So perhaps Aristotle is right that anybody who has given it a moment's thought realizes, or *should* realize, that, when it comes to building character, we become by doing.[13]

But that qualifier about "a moment's thought" introduces a problem. Sad to say, not everyone gives thought to character formation. It is likely that by the early teens children start making character-forming choices, and some may be self-reflective or intellectually curious enough to wonder about character formation and even conclude that choices and overt actions shape character. If they have younger siblings they may well have engaged in some Aristotelian-style education themselves. But probably plenty of early teenagers, old enough to engage in character-forming choices and start building their characters, just have not reached the stage of wondering how virtues and vices are born. One hopes that it is only the rare adult who has never pondered the question of character development, but perhaps this is a misguided hope. The illiterate and

[13] As I noted in Chapter Eight, another sort of evidence may be found in common intuitions about the relationship of character to past actions as it bears on the issue of justified punishment. See my "Retribution, Forgiveness, and the Character Creation Theory of Punishment." *Social Theory and Practice* 33 (2007): 75–103.

poverty-stricken adult whose whole life is spent trying to secure food and shelter may just not have time to wonder about character. And some people, for whatever reason, seem to be congenitally un-self-reflective and stone deaf to the questions philosophy asks. It seems harsh to blame young teenagers, and those to whom the issue just never occurred, for their ignorance. On the other hand, if we say that, due to their ignorance, they are just not responsible for the vast majority of their choices, then we are reverting to a version of the first approach and allowing that there's a lot less moral responsibility to go around than we had first assumed. And those whom we exclude from the fold of the morally responsible are thereby demeaned.

Furthermore, there may be another category of agents who are innocently unaware that choices and actions form character, since they espouse a theory in psychology which holds that there is no character in Aristotle's sense. "Situationism" proposes, based on intriguing experiments, that people are not really the bearers of fixed and consistent virtues and vices.[14] Aristotle would probably argue that this school misinterprets the conclusions of the experiments. Let us suppose that Aristotle is right, but that an honestly misguided Situationist has convinced himself that, not only do choices and actions not produce character, in fact there *is* no character. He sincerely and consistently believes that he does not have a character—even when he is not engaged in the business of professional psychology. Then, when he performs character-forming choices, he is honestly ignorant that he is forming his character. But it seems peculiar to exempt the sincere Situationist from responsibility for the character-determined choices he makes—to demean him by holding that he is far less responsible than the rest of us—assuming that in other relevant respects he is similar to those whom we do hold responsible. Aristotle may be right that most thoughtful and self-reflective people understand, on some level, that their choices affect their characters. Someone who is honestly ignorant of that fact may well be less responsible than someone who is aware of it. But to say that those who are honestly in ignorance about character formation rarely operate as morally responsible agents seems extreme.

The claim that ignorance regarding character formation does not constitute an excuse can be strengthened by looking at an example concerning ignorance regarding the building of *someone else's* character. Consider a busy and ambitious academic, a committed Situationist who is non-culpably convinced that people do not have character (or have so little as to be effectively irrelevant). She knows that parents ought to take care of their children, and so she makes sure her son is

[14] John M. Doris, *Lack of Character* (Cambridge: Cambridge University Press, 2002).

fed and clothed and sent to school, but she does not take thought for his character. (I grant that this example seems rather fantastical. A parent who—whatever she *says* when she is on campus—does not believe in her heart that children have characters that need tending strikes me as a truly odd phenomenon. But for the sake of argument...) So for instance, she is running an errand at the supermarket when he is four, and just as she is leaving she notices that her son has stolen a candy bar. She does not want to be an accessory to stealing, but she also wants to hurry on to a meeting. She does not consider whatever effect going back to pay for the candy bar, or not going back, may have on the character of her son because she does not believe in character. She chooses to go on to her meeting and pretend she did not see the candy bar. And she consistently makes relevantly similar choices, including buying the son whatever he wants, that being less time-consuming than saying "No."

The son could still turn out well, no credit to his mother. But suppose he develops a thoughtless and selfish character, in part because of his mother's choices. Is she not at least partially to blame for that harmful effect, even if she honestly did not believe that her behavior would produce it? (Presumably the son is also blameworthy in that he will have made self-forming choices which embraced and reinforced the bad behavior condoned by his mother.) Parents have a responsibility to try to build good character in their children, and it does not seem to me that non-culpable ignorance, as in the present example, is an adequate excuse for failing to make the effort. If this judgment is correct, then this seems to support the thesis that people have a responsibility to build and maintain good character in themselves and that non-culpable ignorance regarding the fact that we build character through making choices does not undermine that responsibility. But given that some ignorance *does* excuse the agent from responsibility, we need a principled way to show why ignorance that one is forming one's character is not exculpatory.

Revising the Epistemic Requirement

A third option for solving the tracing dilemma involves qualifying the epistemic requirement. Ordinarily it seems that in order for an agent to be held responsible for an act and its consequences the agent must understand (at least in some rough way) that he is engaged in doing the act and (at least in some rough way) what its consequences might be. But this epistemic requirement does not hold for all actions and consequences. On the one hand there is the sort of information required for responsibility. Non-culpable ignorance of facts within this category of information is exculpatory. But there is a different category of information in which ignorance concerning facts about certain sorts of actions and their consequences does not undermine responsibility. (I will grant below that this latter ignorance might *lessen* the degree of an agent's responsibility. But for my purposes it is enough that it does not wholly undermine it.) There is at least one way of making a principled distinction between these two categories of information such that understanding the process of character formation is not required for responsibility.

A developed discussion of the epistemic requisites for free choice lies outside the scope of the present work. (I discussed the issue a little in Chapter Three.) My hope here is to offer some key examples where ignorance regarding actions and consequences does not undermine responsibility, and to propose a principled distinction between this non-exculpatory ignorance and the sort of ignorance which really does render the agent not responsible. Ignorance concerning the fact that one is forming one's character through one's actions will fall on the non-exculpatory side of this distinction. So, if my proposed distinction is plausible, then believing that one is engaged in the activity of character formation may not be required in order for an agent to be responsible for forming his character.

Anselm himself, speaking of his standard example of the fall of Satan, provides an interesting example of ignorance. In order for Satan and his fellow angels to have genuinely open options they cannot know what the results of disobeying God will be. Had they known that they would be cut off from beatitude, disobedience would not have been an option, since no one can will to plunge themselves into misery.[15] But without options there is no choosing *a se* to be good, so God sets up the situation to include the ignorance. Anselm does not even hint that this ignorance should or could undermine responsibility, and surely he is right about this. One possible consequence of a wicked choice—a choice to murder, for example—is punishment. But it would be bizarre to allow ignorance that one was going to be punished to undermine responsibility for the choice and the murder. Suppose you are a criminal mastermind. You have committed several murders and gotten away with it, so you reasonably believe you will not be caught and punished. As it turns out you do get caught. You were non-culpably ignorant of one of the major consequences of your action, but surely "I never dreamed I'd get caught and imprisoned!" does not count *in the slightest* as a mitigating factor. So obviously some instances of ignorance of the consequences of your actions are in no way exculpatory.

But, equally obviously, some instances of ignorance are exculpatory. Suppose you give a friend with a headache some pills from a bottle labeled "Aspirin," and

[15] *De casu diaboli* 23–4.

these pills poison the friend. And suppose you could not have known that the pills were poison and would kill your friend, since you were not aware that someone secretly slipped the pills into the aspirin bottle. In this case you are absolved of any guilt for the death of your friend. In both the case of the punishment and in the case of the poisoning you are ignorant of some of the consequences of your choice and action. But there is a principled distinction to be drawn, and it has to do with what we can refer to as the "content" of your choice. What you are consciously choosing to do, including what you hope to achieve, is the content of your choice. In the poisoning case what you choose to do is give your friend some aspirin to alleviate his headache. Your ignorance is directly related to the content of your choice, since you are innocently unaware that the pills are not aspirin, but rather poison.

In the murder case, what you are choosing to do is commit a murder. What you are innocently unaware of is the fact that (as it happens) a consequence of your choice is that you get punished. But your ignorance is not about some aspect of what you are intending. The punishment is not directly connected to your choice to murder. We can say that belief, or the lack of it, concerning the punishment is extrinsic to the content of the choice. In the Anselmian universe there needs to be a bit of a caveat here. For Anselm, as for Aristotle, you do yourself harm by doing wrong. (This is one facet of the tracing doctrine presently being defended.) This is the inexorable logic of moral activity. So, in one sense, "punishment" of a sort—harm to the agent as a consequence of his bad behavior—*is* an intrinsic consequence of bad choices and deeds. What I mean by "extrinsic to the content of the choice" here is that questions about punishment or other harm to the agent himself are unconnected to the conscious *intentions* of the agent. (In the case of a peculiar agent who deliberately intended to harm himself—if such there be—the harm would be part of the content of the choice.) In the murder example you intend to murder and presumably to achieve various consequences by the murder. Your ignorance that you will be punished is not a lack of understanding about some aspect of what it is you intend. I believe that further examples will support the significance of this distinction between ignorance of aspects of the *content* of choice, as opposed to ignorance involving other phenomena related to choice. And this will help to defend the thought that you can be responsible for forming your character by your choices, even if you do not understand that you are doing so.[16]

[16] It does not follow that ignorance involving the content of your choice—what overt deed you intend and what consequences you hope to achieve from it—is *always* exculpatory. There might be

Prima facie it might seem surprising to hold that if you innocently do not believe that you are doing something, you can nonetheless be held responsible for doing it. Obviously, you should not blame me for stepping on your foot if I, innocently, do not believe that I am doing it. But there is at least one example of an act you engage in where there is likely to be general agreement that you are responsible even if you do not grasp that you are doing the act. That is the *very act of making a choice.* Throughout the present work I have distinguished between the act of your deciding to do something and the subsequent overt deed you chose to do. Suppose you decide to write a check to the Red Cross and then you write it. We may praise you for writing the check. But if we found that the writing was not consequent upon a decision of yours, but rather your limbs were being operated by a magician against your will, and you never intended to write the check, we would retract the praise. Our supposition is that responsibility for the overt act requires a responsibly made choice. Someone with a very different understanding of responsibility than the Anselmian's might doubt this. On the extreme position that ascribing responsibility to an agent means *only* that reward or punishment of that agent by society will have some beneficial consequences, then it might be acceptable to uncouple responsibility for an action from the agent's preceding conscious activity. But someone making this move could jettison the epistemic requirement for responsibility from the outset, and so the tracing problem would not arise.

Most who debate questions of free will and responsibility, whether libertarian, compatibilist, or hard determinist, probably take this connection between choice and overt deed for granted.[17] My suggestion is that for many responsible agents it is often the case that, when they are making what we would agree to be responsible choices, they do not grasp that they are making them. This would seem to be especially true on the Anselmian analysis in which the choice itself is the per-willing of one of two inconsistent desires, such that the choice has no independent ontological status. It is not a new "thing" in the world (see Chapter Four). If the choice has such a "thin" sort of existence, it would not be surprising that we are often not aware, as we are making a choice, that we are actually at the point of making it. But even if most agents are unaware that they are indeed making choices, it seems implausible to hold that they are thereby not responsible for their choices and further, ensuing acts. Insisting, to the contrary,

cases where you ought to have known certain things about the content of the choice and you are blameworthy for your ignorance.

[17] The hard determinist may hold that you do not engage in responsible acts, since you do not make responsible choices, but he does not deny the connection.

that an agent must understand that he is making a responsible choice if he is to be held responsible would radically limit the instances in which we can ascribe responsible agency, a conclusion which demeans people, and so is to be avoided if possible. This ignorance regarding making a choice, which is consistent with responsibility, is different from the ignorance we take to be exculpatory. Like ignorance of punishment, it is ignorance that is extrinsic to the content of the choice. I will argue that ignorance concerning character formation shares this distinguishing feature, and that will support my claim that the epistemic requirement may not apply to responsibility for character formation, and so need not conflict with the tracing thesis.

A concrete case will help make the point that responsible agents may often be in ignorance of making responsible choices, and for present purposes our paradigmatic case of Satan had best be left behind. Here it is better to discuss a choice where the motivations and consequences are more familiar and admit of plausible development. So best to use a human example. Say that an agent is in the torn condition between committing a murder and refraining from doing so. She struggles with her desire to kill and her desire to conform to what she recognizes as the morally right thing to do, refrain. Say that she ends up per-willing to kill. Our present question is this: In order to be responsible for her choice and her subsequent killing, does she need to be aware that she engaged in an act of making a choice of which the effect was her killing? To the philosopher steeped in the free will question it might, at first glance, seem wildly improbable that someone could make a responsible choice without grasping that they are making it. But there are good reasons to hold that people often make choices of which they are unaware.

First, what sort of action do we refer to by "making a choice"? There are various theories of what a choice consists in. We might argue that to be *fully* aware that one is "making a choice," one would need to be aware that one is making a choice under the correct description of "making a choice." That seems far too strong. Perhaps the epistemic requirement is satisfied if the agent has only a very general grasp of what he is doing and its consequences. So an agent could understand *that* he is making a responsible choice which produces his subsequent actions, without understanding what he is doing under any more detailed description. But it is likely that many agents do not have even that little bit of awareness. In the preceding section I listed agents for whom the Aristotelian response to the tracing problem seems to fail—the young teenager, the poverty-stricken worker, and the congenitally un-self-reflective person. These agents do not wonder about character formation, and so it seems likely that they may also be unaware that they engage in the invisible, but effective, actions of making responsible choices. Perhaps a very little Socratic questioning would get them to see that they have in fact engaged in this sort of action. But these agents may never run into Socrates.[18]

The ignorance in question may be more widespread and go deeper than the above suggests. However we flesh out the epistemic requirement, it seems to apply to the agent's understanding *before and/or as* he is doing the act. If you are informed *after* you have given the pills to your friend that they were poison and not aspirin, your excuse that you were ignorant before and during the time you gave him the pills remains viable. So, whatever the Socratic questioning reveals after the fact, if it were universally true that responsibility requires that you need to be aware that you are doing an act as you do it, the responsible agent would need to grasp that he is making a choice at (or possibly before) the time that he is making it. But chances are many agents on many occasions are quite in ignorance of the fact that they are making choices. While there has been a great deal of discussion about what responsible choice consists in, surprisingly little work has been done on the phenomenology of choice—what it is that the agent actually experiences at the time of making a choice.[19] Are we really aware that we are making a responsible choice, *at (and/or before) the moment we make it*?

Sometimes we may be. If you decide right now to raise your left arm and then you raise it, I suppose you have some sort of consciousness of the preceding decision. Perhaps you even say to yourself, "Now I shall raise my arm." But what of the more interesting, morally significant choices that are our concern here? For example, what about our murderer? As she pulls the trigger she may be aware that she has settled on the desire to kill, but is she conscious of engaging in the action referred to as "making a choice"? That seems to require a sort of "standing back" to perceive or assess one's mental states. My suspicion is that in a case such as this the agent would be so concentrated on her struggle between the two options that the "moment of choice" passes without her taking any notice of it. Again, if Anselm's analysis is plausible it is even more likely that someone

[18] There is possibly another class of person who is ignorant of making conscious choices; those who (innocently and sincerely) belong to philosophical and psychological schools of thought which hold that human beings do not really engage in the activity which could recognizably be described as conscious choice. Daniel Wegner may be an example. See *The Illusion of Conscious Will* (Cambridge, MA: Bradford Books, MIT Press, 2002).

[19] Philosophers sometimes mention a "feeling" of freedom, and some work has been done on whether folk intuition regarding the experience of choice leans towards libertarianism or compatibilism, but that does not speak to my question here: What does it actually feel or seem like—if anything—when an agent makes a choice? For the discussion of folk intuition, see Eddy Nahmias, Stephen Morris, Thomas Nadelhoffer, and Jason Turner, "The Phenomenology of Free Will." *Journal of Consciousness Studies* 11 (2004): 162–79.

making a morally significant choice would not detect the action per se, since all it is is the moment at which per-willing one option renders the other no longer viable. And if, as many writers on the subject of free will seem to suppose, the "moment" is effectively instantaneous, then it would be very unlikely that the agent would note it at the time and remember it. (Whether or not it *is* effectively instantaneous remains an open question.)

So we had better not say that an agent must be aware of making a choice in order to be responsible for it, if we hope to avoid severely limiting the number of cases in which we can ascribe responsibility. We can apply the distinction made above between the ignorance which is exculpatory and that which is not, if we focus on the difference between an agent's understanding *that* he is making a choice and his understanding *what* it is he is choosing to do. It is the latter, but not the former, that is required for responsibility. The "what it is he is choosing" is the *content* of the choice. The content covers intentions regarding a variety of things including the overt action that the agent hopes to do, whatever elements he consciously understands to be involved in bringing about the action, and the consequences—the situation, the things, or events—that he hopes to achieve, or that he believes will result from his choice.

In the example where you give your friend the pills, your intending to help your friend get rid of his headache is part of the content of your choice, as is your deciding to go look for the bottle and then handing it to him as the means to your chosen goal. As far as your intentions go, you are succeeding in making the choice to help your friend. But you, non-culpably, are in ignorance that his taking the pills will kill him. Your exculpatory ignorance regards a fact concerning the content of your choice. This seems the proper place to locate the mitigating or excusing factor. We can hold you responsible for a choice, even though you do not know that you are making it, but when we attempt to assess whether you are to be praised or to be blamed, and to what degree, then we look to the content of the choice. The vast majority, perhaps all, of the persuasive examples and arguments in the literature on this subject appeal to ignorance about the content of the choice, and do not touch on ignorance of making the choice itself. So the distinction between the two areas of ignorance seems apt.

You may be responsible for making a choice and for the subsequent actions which are the effects of that choice while in ignorance *that* you are making a choice. Is it plausible, then, to hold that you may be responsible for making a *character-forming* choice and for the subsequent character which is among the effects of that choice while in ignorance *that* you are making a character-forming choice? There are at least two differences between the case of ignorance of making a choice and ignorance of making a *character-forming* choice. First, character formation *can* be part of the content of a choice in that people do sometimes deliberately do things in order to develop their characters. It is unlikely that making a choice *simpliciter* would be part of the content of a choice.[20] This, by itself, does not affect the argument, but it points to a second difference which, prima facie, looks to be problematic for the proposed symmetry between ignorance of making a choice and ignorance of forming one's character by making a choice. The character which is formed by the character-forming choice is an intrinsic *effect* of the choice. In the poisoning case ignorance of the consequences of choice is exculpatory. You chose to give your friend the pills, but you non-culpably did not believe they would kill him. Is ignorance of consequences in general exculpatory? No, the example of the murderer who did not believe he would be punished shows that ignorance of certain sorts of consequences does not excuse or mitigate responsibility. But the punishment example is different from character formation. The punishment is a consequence of your choice only in an indirect way. That is, the murder does not directly cause society to punish you. You might murder and not get caught, or murder and we pardon you for some reason. So the punishment case is different from the case of character formation. With character formation the thesis is that your actions directly produce your character. So the question becomes, does ignorance of the *direct* character-forming *consequences* of your choice free you from responsibility for those consequences, for your character, and for subsequent, character-determined choices? If so, responsible agency is likely to be in short supply.

Happily, although ignorance of the fact that you form your character through your choices is somewhat different from the previous examples about ignorance that you will be punished and ignorance that you are making a choice, the principled distinction suggested by these two examples seems apt in the case of character formation. All three situations share the feature that the ignorance in question is not about the content of your choice. Assuming that you are not deliberately trying to form your character, the goal of character formation does not fall within the content of the choice. It is not *what* you are choosing. The tracing thesis does not come into conflict with the epistemic requirement if we

[20] Perhaps there are cases of choosing to choose. For example, in choosing between what are effectively indiscernibles, you might choose to choose; you can't decide on what you want from the menu, but it's time to order, so you force yourself to choose. The sort of choice the Anselmian is concerned about, though, has to do with morally significant options when you are in the torn condition. In that case it seems unlikely that the agent could step back and choose to choose. But perhaps the ingenious philosopher may be able to come up with cases in which the agent chooses that he should choose—perhaps invoking Frankfurt-style orders of choosing. In these cases character-formation would be more like making a choice *simpliciter*, and that only strengthens my argument.

understand exculpatory ignorance to be ignorance regarding something in the content of choice. Again, we judge the moral value of your choice based on what it is you intend to do, so it is appropriate that the epistemic requirement applies to the content of your choice, but not necessarily to elements extrinsic to the content. You don't need to be aware that you are making a choice, or that punishment will follow upon it, to be responsible for making the choice or engaging in the overt deed it causes. Similarly you don't need to grasp that you are forming your character to be responsible for forming it. This conclusion has the advantage of supporting the traditional understanding of responsibility for character-determined actions without banishing those who are ignorant concerning character formation from the community of morally responsible agents.

There is further evidence in defense of the suggested distinction between ignorance concerning the content of a choice and ignorance concerning character formation. A glance at the doctrine of double effect shows that distinguishing the two categories this way does sometimes operate as a tacit assumption in moral discourse. The doctrine of double effect is a standard principle in Catholic moral thought, though, to my knowledge, it enters into Catholic ethical theory after Anselm's day.[21] The doctrine of double effect comes into play when we are considering an action that is intrinsically permissible, and that is aimed at achieving some proper benefit, but which may have harmful side effects. The doctrine asks us to weigh the benefits of the intended consequences against the costs of whatever unhappy unintended (though anticipated) consequences may ensue. Assuming that the overall benefit of the intended effect outweighs the cost of the unintended effect, the act may be done. Note that the consequences under consideration are consequences *of the overt action which we are intending to do.* This calculation does not consider the consequences on the character formation of the agent. Catholic thought is deeply indebted to Aristotle, and one who subscribes to the doctrine of double effect may well believe that we should consider the unintended harm to the characters of agents engaged in causing collateral harm to achieve an overall benefit. The additional harm to the characters of the agents in question might even tip the scales against doing the proposed action. Nevertheless, the calculation done to decide what overt action is morally permissible, given its unintended side-effects, is separate from the calculation which adds in the harm done to the character of the agent. Thus the doctrine of double effect makes a distinction between consequences regarding the content

[21] Perhaps we might see Anselm's care in noting that we might will some appropriate benefit but not for a morally praiseworthy reason as influential on later, more complex, distinctions; *De Veritate* 12.

of the choice and consequences to the character of the agent. That this distinction is commonly made, at least in some circles, supports the suggestion that we can treat ignorance regarding character formation differently from the way we treat ignorance regarding the content of a choice.

Sometimes we do deliberately try to form our characters. That is, character formation may, for some agents at some times, be part of *what* it is that the agents are choosing, of the content of the choices. In that case it seems plausible to say that the agent is *more* responsible, more praiseworthy or blameworthy, for his character than he would be if he were forming his character in ignorance. He seems to have more control, more input, regarding his character. And this is good news for agents. Ordinarily, when an agent is consciously trying to form his character, he hopes to become a better person.[22] And so the agent who has succeeded in intentionally constructing a better character for himself will bear greater responsibility for his good character and the good behavior it determines, than the agent who has ignorantly constructed a bad character for himself. The latter is still responsible, but less so. Ignorance concerning character formation does affect the *degree* to which an agent can be praiseworthy or blameworthy for a character-determined choice, but it does not negate responsibility. The time-honored tracing thesis remains secure, as does Anselmian libertarianism's hope to defend the thought that we may engage in self-creation.

Conclusion

Almost a millennium ago Anselm of Canterbury took up the task of solving a difficult puzzle. The puzzle is this: While all creation is very good, human beings are unique and best among observable creatures. We owe our status to two special properties. We are rational. And we are also moral agents capable of being responsible for the kind of people we are. That is, we have the amazing capacity for self-creation. But how is this possible? We do not bring ourselves into being. In fact, while we are clever at rearranging what is given in the world, *we do not bring anything at all into being.* How, then, can anything about ourselves be up to us? That is the question that motivates Anselm to construct the theory of free will which I have labeled Anselmian libertarianism.

For some of us, the puzzle hasn't really changed. But here it is well to note that there is a divide among participants in the free will debate—a divide that goes back centuries. For some, the free will issue is essentially the same as it was for

[22] But for a not wildly implausible case of someone trying to make himself a worse person, see the example of Jeff the Jerk in Vargas (2005): 275–6.

Anselm: How can we defend theoretically the very robust aseity required if we human beings are to be the specially valuable things we have traditionally taken ourselves to be? On the other hand, many philosophers hold that all we want from a theory of freedom is that it provide adequate justification to ground our social practice of holding you responsible such that our praising or blaming you will prove beneficial. For those interested mainly in preserving society's practices, the Anselmian concern for defending the elevated status of human agents as self-creators may seem foreign and wrong-headed. Rather than ignoring this divide, I have attempted throughout the present work to insist upon it. So much of the free will debate consists in a clash of intuitions. And it may well be that the differences in intuitions are rooted in differences concerning why "we" care about free will. One contribution of the present work is to serve as an example for how participants in the free will debate may bring their background assumptions into the light. I do not suggest that this method is likely to achieve agreement among philosophers, but at least it should promote clarity and understanding.

In the centuries since Anselm set out his libertarian theory a great deal of work has been done on the free will problem. And the last several decades have seen tremendous strides. One of the great, recent achievements has been clarifying the distinction between, and extensively developing, the two main species of libertarianism; event-causation and agent-causation. Anselm's contribution to this achievement is the suggestion of a new and parsimonious agent-causation. It was the puzzle about human aseity in the universe of classical theism that drove Anselm to make this original move: We can work only with what we are given, and yet *something* must be ultimately up to us. Up to us and *not* up to God, the natural universe, or anything else. Anselm solves the puzzle by insisting on alternative options and the possibility of *a se* choice, which choice does not require any *sui generis* species of causation, nor does it add any new "thing" to the sum of what exists in the universe.

But with aseity and open options at the core of the theory, Anselmian libertarianism is open to three criticisms which are standardly raised against libertarianism. First, if everything about the agent comes from outside, including the motivations which present him with the requisite open options, can it really be the case that that moment of aseity, when the agent opts for this over that, is of such ultimate importance that without it the agent is not free or responsible? The Anselmian responds that it is true that the created agent has very little from himself. Indeed this "minimalism" is one of the virtues of this *parsimonious* agent-causation. In terms of the theory, parsimony is a good thing, and, from the perspective of the created agent, recognizing that he has very little from himself should encourage him to maintain an appropriate humility. But though the scope of the created agent's aseity is very modest, it is enough to ground the remarkable capacity for self-creation.

Second, hasn't Frankfurt demonstrated that agents can have all the freedom that even libertarians want without the (problematic) alternative possibilities that libertarianism insists upon? The Anselmian argues that the Frankfurt-style counterexamples, which have generated such a literature in recent decades, simply cannot connect with Anselmian libertarianism. This is because the insistence upon aseity entails the grounding principle: The truth concerning a libertarian free choice and knowledge about a libertarian free choice come from (or are grounded in or dependent upon) only the actual making of the choice. No one, not even God, can know what S will or would choose until and unless S actually chooses. And if S chooses A it is impossible that S also, simultaneously, choose -A. The Frankfurt-style counterexamples depend upon closing off alternatives by supposing it could be the case that an agent who is going to choose, or is choosing, A can be made to choose -A. But no one, not even God, can bring this about.

And finally there is the luck problem; if there is nothing which explains the choice for this over that beyond the bare fact that the agent made it, doesn't that mean that the choice was just a matter of luck for the agent, not something which arises from his character and thus something for which he can be praised or blamed? The Anselmian, and Anselm himself, grant that it is disquieting to admit that there is no explanation for the agent's choosing this over that. That is a pill that Anselmian libertarianism must swallow. Nonetheless, as numerous libertarians have pointed out, it is peculiar to insist that an agent's deliberately *doing* something is just luck or chance. Moreover, the Anselmian's emphasis on self-creation motivates a new and substantive response. Those who hold that your choices must flow from your character have the relationship between choice and character backwards. What we praise and blame you for is not your character *before* you choose and act, but rather for what you choose and make of yourself by choosing. Perhaps someone who is not interested in self-creation will not find this germane to the luck problem. That shows, again, the importance of understanding the differences among participants in the free will debate concerning just why it is that we care about freedom.

But if the Anselmian is to invoke the importance of self-creation in responding to the luck problem, then he must address a fourth issue, which, for other libertarians, may not be of central concern. Anselmian libertarianism leans heavily on the thought that we can be responsible for our choice-produced characters—that is what being self-creators means—and hence we can be responsible for character-determined choices. But if we do not know, as we are producing our characters by our choices, that we are indeed self-creators, can we truly be responsible? Aristotle's response—that we *ought* to know—is not entirely satisfactory, and so the Anselmian introduces and defends a distinction between a category of ignorance that undermines our responsibility and a category that does not. Being ignorant that we do indeed create our characters by our choices does not absolve us of responsibility for what we make of ourselves.

Though Anselm has been dead for 900 years, his analytic genius is such that his theory has much to offer us today; a parsimonious agent-causation and inspiration for new approaches to the standard problems raised against libertarianism. In the contemporary milieu many philosophers have abandoned the thought that motivated Anselm, that is, that human beings are of special and unique value because of their capacity for self-creation. But perhaps, at least in some instances, this is because philosophers have believed self-creation, and the sort of free choice it requires, to be impossible, wildly unparsimonious, or, at best, seriously implausible or so mysterious that committing to it would be irresponsible. If the argument of the present work is correct, then the thought that we are capable of *a se* choice is defensible. And perhaps we are indeed the wonderful beings that Anselm took us to be.

Bibliography

Adams, Marilyn McCord, and Norman Kretzmann (editors and translators), [William of Ockham's] *Predestination, God's Foreknowledge, and Future Contingents* (New York: Appleton-Century-Crofts, 1969).

Baer, John, "Free Will Requires Determinism." In *Are We Free? Psychology and Free Will* edited by John Baer, James C. Kaufman, and Roy F. Baumeister (Oxford: Oxford University Press, 2008): 304–10.

Baer, John, James C. Kaufman, and Roy F. Baumeister (editors) *Are We Free? Psychology and Free Will* (Oxford: Oxford University Press, 2008).

Baker, Lynne Rudder, "Moral Responsibility without Libertarianism." *Noûs* 40 (2006): 307–30.

Balaguer, Mark, *Free Will as an Open Scientific Problem* (Cambridge, MA: MIT Press, 2010).

Baumeister, Roy, "Free Will, Consciousness, and Cultural Animals." In *Are We Free? Psychology and Free Will* edited by John Baer, James C. Kaufman, and Roy F. Baumeister (Oxford: Oxford University Press, 2008): 65–85.

Beebee, Helen, and Alfred Mele, "Humean Compatibilism." *Mind* 111 (2002): 201–24.

Berofsky, Bernard, "Global Control and Freedom." *Philosophical Studies* 131 (2006): 419–45.

Blumenfeld, David, "Freedom and Mind Control." *American Philosophical Quarterly* 25 (1988): 215–27.

Bostock, David, *Aristotle's Ethics* (Oxford: Oxford University Press, 2000).

Broadie, Sarah, *Ethics with Aristotle* (Oxford: Oxford University Press, 1991).

Campbell, Charles Arthur, "Is 'Freewill' a Pseudo-Problem?" *Mind* 60 (1951): 441–65.

Clarke, Randolph, "Libertarian Views: Critical Survey of Noncausal and Event-causal Accounts of Free Agency." In *The Oxford Handbook of Free Will* (First Edition) edited by Robert Kane (Oxford: Oxford University Press, 2002): 356–85.

Clarke, Randolph, *Libertarian Accounts of Free Will* (Oxford: Oxford University Press, 2003).

Clarke, Roger, "How to Manipulate an Incompatibilistically Free Agent." *American Philosophical Quarterly* 49 (2012): 139–49.

Dennett, Daniel C., *Brainstorm* (Ann Arbor, MI: Bradford Books, 1978).

Dennett, Daniel C., *Elbow Room* (Cambridge, MA: MIT Press, 1984).

Doris, John M., *Lack of Character: Personality and Moral Behavior* (Cambridge: Cambridge University Press, 2002).

Double, Richard, "Libertarianism and Rationality." *The Southern Journal of Philosophy* 26 (1988): 431–9.

Double, Richard, *The Non-reality of Free Will* (Oxford: Oxford University Press, 1991).

Feltz, Adam, Edward T. Cokely, and Thomas Nadelhoffer, "Natural Compatibilism versus Natural Incompatibilism: Back to the Drawing Board." *Mind and Language* 24 (2009): 1–23.

Fischer, John Martin (editor), *God, Foreknowledge, and Freedom* (Stanford, CA: Stanford University Press, 1989).

Fischer, John Martin, "Ultimacy and Alternative Possibilities." *Philosophical Studies* 144 (2009): 15–20.

Fischer, John Martin, and Neal Tognazzi, "The Truth about Tracing." *Noûs* 43 (2009): 531–56.

Fischer, John Martin, Robert Kane, Derk Pereboom, and Manuel Vargas (editors), *Four Views on Free Will* (Oxford: Blackwell Publishing, 2007).

Flint, Thomas P., *Divine Providence: The Molinist Account* (Ithaca, NY: Cornell University Press, 1998).

Frankfurt, Harry, "Alternate Possibilities and Moral Responsibility." *Journal of Philosophy* 66 (1969): 829–39.

Frankfurt, Harry, "Freedom of the Will and the Concept of a Person." *Journal of Philosophy* 68 (1971): 5–20.

Frankfurt, Harry, "Reply to John Martin Fischer." In *Contours of Agency* edited by S. Buss and L. Overton (Cambridge: Cambridge University Press, 2002): 27–31.

Freddoso, Alfred J., *On Divine Foreknowledge* (Ithaca, NY: Cornell University Press, 1988).

Ginet, Carl, *On Action* (Cambridge: Cambridge University Press, 1990).

Ginet, Carl, "In Defense of the Principle of Alternative Possibilities: Why I Don't Find Frankfurt's Argument Convincing," *Philosophical Perspectives* 10 (1996): 403–17.

Glock, Hans-Johann, "Animal Agency" in *A Companion to the Philosophy of Action* edited by Timothy O'Connor and Constantine Sandis (Chichester, UK: Wiley-Blackwell, 2010): 384–92.

Green, Jeffrey and Katherin Rogers, "Time, Foreknowledge, and Alternative Possibilities." *Religious Studies* 48 (2012): 151–64.

Griffith, Meghan, "Why Agent-Causal Actions Are not Lucky." *American Philosophical Quarterly* 47 (2010): 43–56.

Hasker, William, *God, Time, and Knowledge* (Ithaca, NY: Cornell University Press, 1989).

Hobart, R. E., "Free Will as Involving Determination and Inconceivable without It." *Mind* 43 (1934): 1–27.

Hodgson, David, "Quantum Physics, Consciousness, and Free Will." In *The Oxford Handbook of Free Will* (First Edition) edited by Robert Kane (Oxford: Oxford University Press, 2002): 85–110.

Howard, George S., "Whose Will? How Free?" In *Are We Free? Psychology and Free Will* edited by John Baer, James C. Kaufman, and Roy F. Baumeister (Oxford: Oxford University Press, 2008): 260–74.

Hunt, David, "Freedom, Foreknowledge, and Frankfurt." In *Moral Responsibility and Alternative Possibilities* edited by David Widerker and Michael McKenna (Farnham, UK: Ashgate Publishing Limited, 2003): 157–83.

Hunt, David, "Moral Responsibility and Buffered Alternatives." *Midwest Studies in Philosophy* 29 (2005): 126–45.

Hunt, David, "Black the Libertarian." *Acta Analytica* 22 (2007): 3–15.

Irwin, Terence Henry, "Reason and Responsibility in Aristotle." In *Essays on Aristotle's Ethics* edited by Amelie Oksenberg Rorty (Berkeley, CA: University of California Press, 1980): 117–55.

James, William, "The Dilemma of Determinism." *Unitarian Review* 22 (1884): 193–224.

Kane, Robert, *The Significance of Free Will* (Oxford: Oxford University Press, 1996).

Kane, Robert, "Responsibility, Indeterminism and Frankfurt-style Cases: A Reply to Mele and Robb." In Widerker and McKenna (2003): 91–105.

Kane, Robert, "Libertarianism." In *Four Views on Free Will* edited by John Martin Fischer, Robert Kane, Derk Pereboom, and Manuel Vargas (Oxford: Blackwell Publishing, 2007): 5–43.

Kane, Robert, "Three Freedoms, Free Will and Self-formation." In *Essays on Free Will and Moral Responsibility* edited by Nick Takakis and Daniel Cohen (Newcastle on Tyne: Cambridge Scholars Publishing, 2008): 142–62.

Kane, Robert, "Libertarianism." *Philosophical Studies* 144 (2009): 35–44.

Klein, Colin, "Philosophical Issues in Neuroimaging." *Philosophy Compass* 5 (2010): 186–98.

Libet, Benjamin, "Do We Have Free Will?" In *The Oxford Handbook of Free Will* (First Edition) edited by Robert Kane (Oxford: Oxford University Press, 2002): 551–64.

Loewer, Barry, "Freedom from Physics: Quantum Mechanics and Free Will." *Philosophical Topics* 24 (1996): 91–112.

Lowe, E. J., *Locke on Human Understanding* (London: Routledge, 1995).

Lowe, E. J., *Personal Agency* (Oxford: Oxford University Press, 2008).

McCann, Hugh, "Divine Sovereignty and the Freedom of the Will." *Faith and Philosophy* 12 (1995): 582–9.

McCann, Hugh, *The Works of Agency: On Human Action, Will, and Freedom* (Ithaca: Cornell University Press, 1998).

McCann, Hugh, "Sovereignty and Freedom: A Reply to Rowe." *Faith and Philosophy* 18 (2001): 110–16.

McCann, Hugh, "The Author of Sin?" *Faith and Philosophy* 22 (2005): 144–59.

McCann, Hugh, "God, Sin, and Rogers on Anselm: A Reply." *Faith and Philosophy* 26 (2009): 420–31.

McCann, Hugh, *Creation and the Sovereignty of God* (Bloomington and Indianapolis: Indiana University Press, 2012).

McKenna, Michael, "A Hard-Line Reply to Pereboom's Four-Case Manipulation Argument." *Philosophy and Phenomenological Research* 77 (2008): 142–59.

Mele, Alfred, *Autonomous Agents* (Oxford: Oxford University Press, 1995).

Mele, Alfred, "Introduction." In *The Philosophy of Action* edited by Alfred Mele (Oxford: Oxford University Press, 1997): 1–26.

Mele, Alfred R., "Ultimate Responsibility and Dumb Luck." *Social Philosophy and Policy* 16 (1999): 274–93.

Mele, Alfred, *Free Will and Luck* (Oxford: Oxford University Press, 2006).

Mele, Alfred, *Effective Intentions* (Oxford: Oxford University Press, 2009a).

Mele, Alfred, "Moral Responsibility and Agents' Histories." *Philosophical Studies* 142 (2009b): 161–81.

Mele, Alfred, "Moral Responsibility and the Continuation Problem." *Philosophical Studies* 162 (2013): 237–55.

Mele, Alfred, and David Robb, "Rescuing Frankfurt-Style Cases." *Philosophical Review* 107 (1998): 97–112.

Mele, Alfred, and David Robb, "Bbs, Magnets and Seesaws: The Metaphysics of Frankfurt-style Cases." In *Moral Responsibility and Alternative Possibilities* edited by David Widerker and Michael McKenna (Farnham, UK: Ashgate Publishing Limited, 2003): 127–38.

Merricks, Trenton, "Truth and Freedom." *Philosophical Review* 118 (2009): 29–57.

Miller, William R. and David J. Atencio, "Free Will as a Proportion of Variance." In *Are We Free? Psychology and Free Will* edited by John Baer, James C. Kaufman, and Roy F. Baumeister (Oxford: Oxford University Press, 2008): 275–95.

Myers, David G., "Determined and Free." In *Are We Free? Psychology and Free Will* edited by John Baer, James C. Kaufman, and Roy F. Baumeister (Oxford: Oxford University Press, 2008): 32–43.

Nagle, Thomas, "Moral Luck." In *Mortal Questions* (New York: Cambridge University Press, 1979): 24–38.

Nahmias, Eddy, "Scientific Challenges to Free Will." In *A Companion to the Philosophy of Action* edited by Timothy O'Connor and Constantine Sandis (Chichester, UK: Wiley-Blackwell, 2010): 345–56.

Nahmias, Eddy, Stephen Morris, Thomas Nadelhoffer, and Jason Turner, "The Phenomenology of Free Will." *Journal of Consciousness Studies* 11 (2004): 162–79.

New, Christopher, "Time and Punishment." *Analysis* 52 (1992): 35–40.

Newsome, William T. "Human Freedom and 'Emergence'." In *Downward Causation and the Neurobiology of Free Will* edited by Nancey Murphy, George F. R. Ellis, and Timothy O'Connor (Berlin: Springer-Verlag, 2009): 53–62.

Nichols, Shaun, "How Can Psychology Contribute to the Free Will Debate?" In *Are We Free? Psychology and Free Will* edited by John Baer, James C. Kaufman, and Roy F. Baumeister (Oxford: Oxford University Press, 2008): 10–31.

O'Connor, Timothy, *Persons and Causes* (Oxford: Oxford University Press, 2000).

O'Connor, Timothy, "Degrees of Freedom." *Philosophical Explorations* 12 (2009): 119–25.

Pakaluk, Michael, *Aristotle's Nicomachean Ethics* (Cambridge: Cambridge University Press, 2005).

Pereboom, Derk, "Determinism al Dente." *Noûs* 29 (1995): 21–45.

Pereboom, Derk, *Living Without Free Will* (Cambridge: Cambridge University Press, 2001).

Pereboom, Derk, "A Hard-line Reply to the Multiple-Case Manipulation Argument." *Philosophy and Phenomenological Research* 77 (2008): 160–70.

Pereboom, Derk, "Hard Incompatibilism and its Rivals." *Philosophical Studies* 144 (2009): 21–33.

Perszyk, Ken (editor), *Molinism: The Contemporary Debate* (Oxford: Oxford University Press, 2011).

Pink, Thomas, "Freedom and Action Without Causation." In *The Oxford Handbook of Free Will* (Second Edition) edited by Robert Kane (Oxford: Oxford University Press, 2011): 349–65.

Plantinga, Alvin, "Supralapsarianism, or 'O Felix Culpa'." In *Christian Faith and the Problem of Evil* edited by Peter van Inwagen (Grand Rapids, MI: William B. Eerdmans Publishing Company, 2004): 1–25.

Roberts, Jean, "Aristotle on Responsibility for Action and Character." *Ancient Philosophy* 9 (1989): 23–36.

Rogers, Katherin, *Perfect Being Theology* (Edinburgh: Edinburgh University Press, 2000).
Rogers, Katherin, "What's Wrong with Occasionalism?" *American Catholic Philosophical Quarterly* 75 (2001): 345–69.
Rogers, Katherin, "Does God Cause Sin? Anselm of Canterbury versus Jonathan Edwards on Human Freedom and Divine Sovereignty." *Faith and Philosophy* 20 (2003): 371–8.
Rogers, Katherin, "God Is not the Author of Sin: An Anselmian Response to McCann." *Faith and Philosophy* 24 (2007a): 300–10.
Rogers, Katherin, "Retribution, Forgiveness, and the Character Creation Theory of Punishment." *Social Theory and Practice* 33 (2007b): 75–103.
Rogers, Katherin, *Anselm on Freedom* (Oxford: Oxford University Press, 2008).
Rogers, Katherin, "Anselm Against McCann on God and Sin: Further Discussion." *Faith and Philosophy* 28 (2011): 397–415.
Rogers, Katherin, "Anselm on the Ontological Status of Choice." *International Philosophical Quarterly* 52 (2012a): 183–98.
Rogers, Katherin, "The Divine Controller Argument for Incompatibilism." *Faith and Philosophy* 29 (2012b): 275–94.
Rogers, Katherin, "Freedom, Science, and Religion." In *Scientific Approaches to the Philosophy of Religion* edited by Yujin Nagasawa and Erik J. Wielenberg (London: Palgrave Macmillan, 2012c): 237–54.
Rogers, Katherin, "Anselm on Self-reflection in Theory and Practice." *The Saint Anselm Journal* 9 (2013) <http://www.anselm.edu/Documents/Institute%20for%20Saint%20Anselm%20Studies/Fall%202013/Anselm%20on%20Self-reflection%20in%20Theory%20and%20Practice.pdf>.
Shanley, Brian, O. P., "Beyond Libertarianism and Compatibilism: Thomas Aquinas on Created Freedom." In *Freedom and the Human Person* edited by Richard Velkley (Washington, DC: Catholic University of America Press, 2007): 70–89.
Shariff, Azim F., Johathan Schooler, and Kathleen D. Vohs, "The Hazards of Claiming to Have Solved the Hard Problem of Free Will." In *Are We Free? Psychology and Free Will* edited by John Baer, James C. Kaufman, and Roy F. Baumeister (Oxford: Oxford University Press, 2008): 181–204.
Sher, George, *In Praise of Blame* (Oxford: Oxford University Press, 2006).
Smilansky, Saul, *Free Will as Illusion* (Oxford: Oxford University Press, 2000).
Smilansky, Saul, "Free Will: Some Bad News." In *Action, Ethics, and Responsibility* edited by Joseph Keim Campbell, Michael O'Rourke, and Harry S. Silverstein (Cambridge, MA: A Bradford Book, MIT Press, 2010): 187–201.
Sommers, Tamler, "Experimental Philosophy and Free Will." *Philosophy Compass* 5 (2010): 199–212.
Steward, Helen, *A Metaphysics for Freedom* (Oxford: Oxford University Press, 2012).
Strawson, Galen, "The Bounds of Freedom." In *The Oxford Handbook of Free Will* (First Edition) edited by Robert Kane (Oxford: Oxford University Press, 2002): 441–60.
Stump, Eleonore, "Alternative Possibilities and Moral Responsibility: The Flicker of Freedom." *The Journal of Ethics* 3 (1999): 299–324.
Stump, Eleonore, "Augustine on Free Will." In *The Cambridge Companion to Augustine*, edited by Eleonore Stump and Norman Kretzmann (Cambridge: Cambridge University Press, 2001): 124–47.

Stump, Eleonore, "Moral Responsibility without Alternative Possibilities." In *Moral Responsibility and Alternative Possibilities* edited by David Widerker and Michael McKenna (Farnham, UK: Ashgate Publishing Limited, 2003): 139–58.

Timpe, Kevin, "Free Will: Alternatives and Sources." In *Philosophy Through Science Fiction* edited by Ryan Nichols, Fred Miller, and Nicholas D. Smith (New York: Routledge, 2008a): 397–408.

Timpe, Kevin, *Free Will: Sourcehood and Its Alternatives* (London: Continuum, 2008b).

Timpe, Kevin, "Tracing and the Epistemic Condition on Moral Responsibility." *Modern Schoolman* 88 (2011): 5–28.

Turner, Jason, "The Incompatibility of Free Will and Naturalism." *Australasian Journal of Philosophy* 87 (2009): 565–87.

Usher, Marius, "Control, Choice, and the Convergence/Divergence Dynamics: A Compatibilistic Probabilistic Theory of Free Will." *The Journal of Philosophy* 103 (2006): 188–213.

van Inwagen, Peter, *An Essay on Free Will* (Oxford: Clarendon Press, 1983).

van Inwagen, Peter, "When Is the Will Free?" In *Philosophical Perspectives 3: Philosophy of Mind and Action Theory* edited by James Tomberlin (Atascadero, CA: Ridgeview Publishing, 1989): 399–422.

van Inwagen, Peter, "Free Will Remains a Mystery." In *Philosophical Perspectives 14: Action and Freedom* edited by James E. Tomberline (Oxford: Blackwell Publishing, 2000): 1–19.

van Inwagen, Peter, "Genes, Statistics, and Desert." In *Genetics and Criminal Behavior* edited by David Wasserman and Robert Wachbroit (Cambridge: Cambridge University Press, 2001): 225–42.

Vargas, Manuel, "The Trouble with Tracing." In *Midwest Studies in Philosophy 26: Free Will and Moral Responsibility* edited by Peter A. French, Howard K. Wettstein, John Martin Fischer (Guest Editor) (Oxford: Blackwell Publishing, 2005): 269–91.

Waller, Bruce, "Free Will Gone Out of Control." *Behaviorism* 16 (1988): 149–57.

Watson, Gary, "Free Action and Free Will." *Mind* 96 (1987): 145–72.

Wegner, Daniel, *The Illusion of Conscious Will* (Cambridge, MA: MIT Press, 2002).

Wegner, Daniel, "Self Is Magic." In *Are We Free? Psychology and Free Will* edited by John Baer, James C. Kaufman, and Roy F. Baumeister (Oxford: Oxford University Press, 2008): 226–47.

Widerker, David, "Libertarianism and Frankfurt's Attack on the Principle of Alternative Possibilities." *Philosophical Review* 104 (1995): 247–61.

Widerker, David, "Blameworthiness and Frankfurt's Argument Against the Principle of Alternative Possibilities." In *Moral Responsibility and Alternative Possibilities* edited by David Widerker and Michael McKenna (Hants, England: Ashgate Publishing Limited, 2003): 53–73.

Widerker, David, and Michael McKenna, editors, *Moral Responsibility and Alternative Possibilities* (Hants, England: Ashgate Publishing Limited, 2003).

Wolf, Susan, *Freedom within Reason* (New York: Oxford University Press, 1990).

Yagisawa, Takashi, "Possible Objects" in the *Stanford Encyclopedia of Philosophy* (2009).

Zimmerman, Dean, "An Anti-Molinist Replies." In *Molinism: The Contemporary Debate*, edited by Ken Perszyk (Oxford: Oxford University Press, 2011): 163–86.

Index

THE KING'S COLLEGE LIBRARY
3 0100 00024 3941